Magale Library
Southern Arkansas University
Magnolia, AR 71753

CHEMICAL HAZARDS
at Water and Wastewater Treatment Plants

RUTH ANN BUZZI

LEWIS PUBLISHERS
Boca Raton   Ann Arbor   London   Tokyo

Library of Congress Cataloging-in-Publication Data

Buzzi, Ruth Ann.
  Chemical hazards at water and wastewater treatment plants/Ruth Ann Buzzi.
    p.   cm.
  Includes bibliographical references and index.
  ISBN 0-87371-491-1
  1. Water treatment plants—Safety measures. 2. Sewage disposal plants—Safety measures. 3. Hazardous substances—Safety measures. I. Title.
TD434.B89   1992
628.1'028'9—dc20   92-2740
                                                                        CIP

COPYRIGHT © 1992 by LEWIS PUBLISHERS
ALL RIGHTS RESERVED

This book represents information obtained from authentic and highly regarded sources. Reprinted material is quoted with permission, and sources are indicated. A wide variety of references are listed. Every reasonable effort has been made to give reliable data and information, but the author and the publisher cannot assume responsibility for the validity of all materials or for the consequences of their use.

Neither this book nor any part may be reproduced or transmitted in any form or by any means, electronic or mechanical, including photocopying, microfilming, and recording, or by any information storage and retrieval system, without permission in writing from the publisher.

LEWIS PUBLISHERS
121 South Main Street, Chelsea, MI 48118

PRINTED IN THE UNITED STATES OF AMERICA
1 2 3 4 5 6 7 8 9 0
Printed on acid-free paper

To the men and women who safeguard our water resources.

Ruth Ann Buzzi was born and raised in Akron, OH. She graduated from Heidelberg College in Tiffin, OH with a B.S. in biology. From 1971 to 1977 she worked on a variety of projects in water pollution control research. In 1979 she graduated from Florida Institute of Technology in Melbourne, FL with a M.S. degree in environmental science. Both in 1971 and 1978 she was listed in *Who's Who of American Colleges and Universities* for her activities in the environmental field. She taught environmental health and wastewater courses for eight years: first at the Jensen Beach FIT branch campus in Jensen Beach, FL, then at Stevens Institute of Technology in Hoboken, NJ. Since 1987 she has been working for the Summit County Environmental Services at the Fishcreek Sewage Treatment Plant in Stow, OH. She has researched and written material for the Cuyahoga River Remedial Action Plan Board and has published articles on chemical safety in the *Bench Sheet* and the *Operator's Forum* magazines. She also holds a Class III Wastewater Operator license and a Class III Wastewater Analyst license with the state of Ohio.

PREFACE

The sewage and water treatment fields are two of the most hazardous professions known. Not only are workers exposed to the usual mechanical hazards, but they are also exposed to a number of chemical and pathological dangers. Given the broad range of wastes, reagents, and diseases encountered by water quality personnel, it is clear that proper safety training is paramount. This work is dedicated to the belief that a well-trained chemist or operator is a safe worker.

This work describes most of the common dangers encountered by the worker on his or her rounds. Since gases, both as byproducts of a process and as reagents used in a process, are encountered most frequently in the workplace, they are covered in great detail. Common process-generated gases are hydrogen sulfide, methane, carbon dioxide, and monoxide. Gases most frequently used in treatment are chlorine, ozone, and sulfur dioxide. All of these are intensely irritating and quick acting. They require thorough understanding of their chemistry, and the utmost care in their handling.

Heavy metals must be considered from two standpoints: they are systemic poisons, and some metals are also mutagenic or teratogenetic in their actions. Therefore, a whole chapter is devoted to birth defect-inducing agents.

Few organic chemicals are encountered in most treatment plants, yet because of their great addictive and central nervous system potential, they must be dealt with in a careful fashion. Certain organic pesticides are also cumulative over time, and their effects are additive with repeated exposure.

Strong acids, bases, and oxidizing agents have great burning and explosive capacities. Water and wastewater treatment relies heavily on pH-adjusting chemicals and oxidizers. If these agents are handled incorrectly, they can cause severe injury, fires, or explosions. Cleanup procedures are just as important as first aid.

Of all hazardous fields except medicine, only wastewater treatment deals with such a horrific variety of diseases. Wastewater personnel are exposed to typhoid, cholera, polio, hepatitis, and certain parasitic diseases. At all times a water pollution worker must practice proper hygiene and keep his or her immunizations up to date.

No manual would be complete without a plan for safety training and emergencies. The last chapter will also contain a list of references that no workplace should be without, including a list of national and local contacts for difficult and unusual problems.

I wish to express gratitude for help in composing this work. I thank my parents, who supported this endeavor and showed great patience with my moods during this work. I thank Dr. Francis J. Waickman, an environmental physician, and Dr. Gene Swanson of Akron Children's Hospital for their expertise in the area of toxicology. I would also like to acknowledge the fine work of Dr. Robert Felter and David J. Uehlein of the Akron Regional Poison Center in reviewing this work. Without their suggestions this work would have been impossible. It would have been impossible to assemble this work without the help of the Kent State University Library reference staff and the wonderful librarians of the Summit County Library at the Stow Branch. The Stow librarians did everything possible to aid my search for important references. Thanks goes to the men and women of Plant 25 in Stow for suggesting topics of safety that would be useful to the pollution worker. Much thanks go to the Water Pollution Control Federation magazines *Bench Sheet* and *Operations Forum* for giving me a start in the career of technical writing. And thanks goes to a fellow writer, Lynne Benson, for guiding me through the intricacies of writing a first book.

CONTENTS

1. Introduction 1

2. Chemical Hazards — Pulmonary Agents 5
   2.1 Asphyxiants 5
   2.2 Respiratory Irritants and Oxidants 11

3. Chemical Hazards —Heavy Metals 23
   3.1 Sources of Contact 23
   3.2 Special Dangers in the Field: Coagulants and Algicides 33

4. Chemical Hazards — Organics 41
   4.1 Sources of Contact 41
   4.2 Health Effects of Organics 42

5. Caustics and Corrosives 61
   5.1 Acids and Bases 61
   5.2 Some Special Precautions 62
   5.3 A Warning about Hydrofluoric Acid and its Byproducts 64
   5.4 A Word of Caution with Caustics 67

6. Dangers of Strong Oxidants and Reduction Agents 73
   6.1 Dangers in the Plant 73
   6.2 Dangers in the Laboratory 75

7. Microbial Hazards in Water 85
   7.1 The Viruses 86
   7.2 Bacteria 87
   7.3 The Protozoa 88
   7.4 Fungi 88
   7.5 Helminths and Other Higher Parasites 89
   7.6 Precautions 89
   7.7 Special Precautions 90

8. Human Reproduction and Chemical Exposure 93
   8.1 Some Important Terms 94
   8.2 Factors of Exposure 95
   8.3 Hazards in the Lab 96

9 Setting Up a Safety Program 107
   9.1 Being Prepared 108
   9.2 Where to Look For Needed Information and Care 109
   9.3 The Plan 112

Appendix A — A Word About Toxicology	115
Glossary	119
Bibliography	125
Index	129

LIST OF FIGURES

2.1	SCBA Setup and Storage Cabinent	7
2.2	Cartridge Respirators	14
3.1	A Mercury Vapor Trap	26
3.2	A Safe Mercury Vapor Setup for Analysis	26
3.3	A Safe Setup for TKN Analyis	28
3.4	Safety with Chromate Salts	32
4.1	Organophosphate Structure	43
4.2	Methyl Carbamates	46
4.3	Pyrethrum	46
4.4	Bipyridyl Compounds	47
4.5	Terpenes	50
4.6	Toulene and Related Compounds	50
4.7	Ketones	52
4.8	Ethers	53
4.9	The Alkanes	54
4.10	Three Commonly Used Alcohols	54
4.11	Halogenated Hydrocarbons	56
4.12	Phenols	56
4.13	Benzene	58
4.14	Formaldehyde	58
5.1	Proper Spill Control	63
5.2	Proper Protection	64
6.1	Types of Fire Extinguishers	78
6.2	Summary of Classes of Fires	79
6.3	A Fire Station for a Hazardous Area	79
9.1	A Safety and First Aid Station	110
9.2	Some Well-Stocked First Aid Kits	111
9.3	A Typical Accident Form	113

LIST OF TABLES

2.1	A Short Summary and Guide for the Handling of Gases	20
3.1	Metals and their Properties	37
4.1	Dangerous Organic Compounds	42
5.1	Safe Handling of the Acids	69
5.2	Safe Handling of the Bases	70

1
Introduction

It has been said by the Water Pollution Control Federation and the Operator's Training Committee of Ohio that jobs in the areas of water and wastewater treatment are some of the most hazardous professions known.[1] Certainly, no other profession has the combination of biological and chemical risks found in this field. Not a year passes that one does not hear of a major injury or death somewhere in the United States, related to activities in the area of water reclamation. The employer must be aware of such risks and take steps to see that his or her workers are properly trained and protected. It is the purpose of this work to focus primarily on the chemical and some of the biological dangers encountered in these vocations.

For the employer and the employee to be safe and effective in their work, they must have at least a basic knowledge of toxicology and physiology. First aid classes and classes in CPR should be offered, from time to time, at the place of work. But training is not enough. Many large companies have their own medical staff on the worksite. Such a staff includes at least one specialist in environmental medicine. Unfortunately, many sewage and water treatment plants are small and on restricted budgets. The next best solution for these workplaces, would be to establish contact with the toxicology department of the local hospital. The hospital should be made aware of the workers in the area and the agents to which the workers might be exposed. A library should be placed in a central location at work and stocked with at least one good chemical index, a lab safety manual, and a comprehensive first aid book. In addition to such a library and training, detailed records of work injuries, actions taken, and outcomes should be kept by the superintendent. The phone numbers of the fire department, disaster services hotline, and the area's poison center should be posted at several prominent areas in the workplace. At least two persons per shift should take it upon themselves to become aware of the various dangers and to learn appropriate measures against such dangers.

All chemicals that are shipped today come with information sheets dealing with reactivity and toxicity. Separate Material Safety Data Sheet (MSDS) files must be established for these chemicals. The files must be updated at least once a year. Such files must be readily available for any worker who desires information on a certain agent.

Even with the best safety programs, accidents will occur from time to time. What is done in the first minutes of an incident could spell the difference between death and disability, or life and health. Many chemicals are quite specific in their actions and will make their acute effects felt within minutes of exposure. First aid measures for such exposures are equally specific. The wrong first aid can actually intensify an injury. Established emergency procedures for chemical and biological exposures are important. They are the first line of defense in the rescue of a worker from injury. A proper response reduces the time between injury and aid, and neutralizes the worst effects of the exposure.

Although procedures vary for different chemicals, there are some first actions to be taken. There are three basic priorities when dealing with such injuries: remove, resuscitate, and neutralize.[2] In the removal of a stricken worker from danger, THE RESCUER MUST TAKE STEPS TO PROTECT HIMSELF. Would-be rescuers have died themselves simply because they did not take proper steps for self-protection. Protective clothing and a breathing apparatus should be kept ready at all times in areas where dangers are known to exist. Once the injured worker is removed, one must then decide which action is to be initiated. If the person is breathing and conscious, then removing a chemical from the skin or lungs may be the first priority. However, if the person is not conscious or breathing, then resuscitation is of the utmost importance. The ideal response to an emergency would allow one person to work on flushing or neutralizing the poison and another person to work on keeping the victim breathing and from collapsing and out of shock. Even as first aid is administered, a person should call 911 for an ambulance: No one else should attempt to drive a person to the hospital. Even if the injured party protests that he or she feels better, workplace policy should demand that he or she goes to a physician for observation. Many toxins are delayed in their more serious effects.

Neutralization is the first line of defense in first aid, and it can be a tricky business. Organic solvents, barrier creams, or neutralizing solutions should not be used if there is the slightest doubt as to the nature of the poison. Flushing a wound or burn with copious amounts of water is the best first line of defense. One should not be shy in removing clothing that may be contaminated. Very often clothing holds a toxin within its fabric. After the injured area is completely washed with plain, cool water, it should be wrapped in clean, wet cloth or gauze. Make sure that a full description of the poison and its chemical structure is called ahead to the emergency room. This saves the attending physician vital time because he knows exactly what kind of agent caused the injury.

With very large spills there is also the problem of environmental contamination. Remember, all large chemical spills must be reported to a local disaster services hotline. For our Summit County, OH, this number is (216)-379-2558. If possible, such spills must be contained or neutralized. But remember, such action should not be attempted by work personnel unless they are very familiar with the chemistry of the toxicant. There are agencies that specialize in the containment and neutralization of spills.

Keeping these simple rules in mind can markedly increase an injured person's chance of complete recovery. Keeping detailed records of injuries and the actions taken can aid in the safe and efficient treatment of future injuries.

REFERENCES

1. Joint Committee of the Water Pollution Control Federation and American Society of Civil Engineers. 1982. *Wastewater Treatment Plant Design*. Lancaster Press Inc., Lancaster, PA.
2. Proctor, N.H., and J.P. Hughes. 1978. *Chemical Hazards of the Workplace*. J.B. Lippencott, Philadelphia.

2
Chemical Hazards — Pulmonary Agents

2.1 ASPHYXIANTS

Although the chemists in water and wastewater labs are often exposed to harmful atmospheres, the problem of respiratory toxins is really most critical in operations and field work. In working in manholes and wet wells, as well as deep excavations for groundwater, the worker is exposed to a number of asphyxiants. An asphyxiant is an agent that can either compete actively for the body's hemoglobin, crowd out oxygen in the surrounding atmosphere, or damage the breathing centers of the central nervous system (CNS). All of these classes of agents are a product of biological decay or a byproduct of some reclamation process.

There are three agents generated in the course of biological activity. These are carbon dioxide, methane, and hydrogen sulfide. Carbon dioxide and methane simply collect in low-lying pockets where air circulation is absent, and crowd out oxygen from that area. Their toxic effects can be reversed if the worker is removed and resuscitated in time. Hydrogen sulfide is another matter. This agent not only crowds oxygen from the atmosphere, but it can actually damage the breathing centers of the body. Each agent will be reviewed individually, and their safe handling will be discussed.

Carbon Dioxide — CO_2

There is not a biological process in sewage treatment that does not generate some carbon dioxide. Normally, this is not a problem, as the gas itself is not toxic. However, should a ventilation system break down, it could become a potentially dangerous gas. This is particularly true in northern climates where sewage processes are kept under shelter due to harsh conditions in winter.

It has been noticed by some operators that when a shed housing a Rotating Biological Contactor (RBC) system loses the use of its fans, a match will not stay lit in the shed. RBCs are a cross between trickling filters and activated sludge systems. They require a great deal of oxygen and give off much carbon dioxide. Oxygen is passively adsorbed by the biological film as it makes its pass on the wheel out of the water and into the air. Two things are happening here: atmospheric oxygen is being depleted, and carbon dioxide is being released.

Carbon dioxide is tasteless, odorless, and colorless. The body's nerve centers, luckily, don't respond to the lack of oxygen in the bloodstream, in order to control breathing. They respond to the increase of CO_2 in the blood.[1] So it is possible for the operator to feel some panting distress and remove himself or herself from the situation. The real danger is that the operator is not paying attention to his or her body or surroundings. The signal for the need to breathe can be subtle, and a worker can already be incapacitated before he or she realizes anything is wrong. An important rule of thumb is that if the ventilation system has broken down in the shed enclosing a process, then the worker should check the atmosphere with a gas detector. (However, one should not use the match trick, in case of the possibility of methane.) In order for an area to be safe for entry, the CO_2 content should be below 0.5% or 5000 ppm, and the oxygen level above 18% or pO^2 135 mm.[2,3] Even in a good oxygen-rich atmosphere, as little as 2% CO_2 can cause breathing and heart rhythm problems. Between 7% and 10% can render a person unconscious in a matter of minutes.[2]

A buddy system should be used when entering suspect areas. One or two people should be tenders on the lookout for trouble. NEVER enter a suspect area alone. Use a gas-ratio analyzer, suited for detecting oxygen, carbon dioxide, and methane, to check the area first. The author has personally seen the results of careless entry. An operator went into a manhole alone, without the benefit of an analyzer or breathing apparatus to read a meter. He was not missed for some minutes, except for his absence from a meeting. Fortunately, the superintendent, who knew the man's schedule, immediately donned harness and breathing gear, located the manhole, and got the man out. As it was, the operator spent an uncomfortable 3 days in the hospital and another week out of work. He was lucky to be alive.

If it has been established that the area is high in carbon dioxide or any other asphyxiant, then a source of forced ventilation must be provided before entry for work. If an area must be entered before ventilation is complete, then one must use a self-contained breathing apparatus (SCBA). Particulate and cartridge respirators are useless for this kind of work, as their action is one of filtration and not oxygen provision (see Figure 2.1).

CHEMICAL HAZARDS — PULMONARY AGENTS

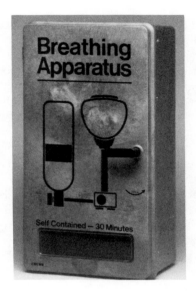

A

B

Figure 2.1 SCBA SETUP and STORAGE CABINET. (A) It is very important to have a highly visible storage area within easy access. (B) The air supply should be at least 15 minutes (30 to 40 minutes is preferred). Make sure all workers are trained in the SCBA's proper use. (Photo courtesy of Lab Safety Supply Inc., Janesville, WI.)

If a person is found unconscious due to simple carbon dioxide exposure, then removal and resuscitation are the first order of business. Sometimes just removal from the area is enough to revive the person. If a person is not breathing, immediately check for a heartbeat, clear the airway, and begin artificial resuscitation or CPR as needed. An ambulance should be called, and the person should be taken to a hospital for observation. Oxygen should only be administered by trained personnel.

Acute effects are only a small part of the toxic picture. Even with as little as 2% CO_2 in an otherwise adequately oxygenated area, workers may complain of headaches, dizziness, and shortness of breath. Good cross-ventilation should be engineered in every building, and care should be taken to see that the CO_2 never gets above 0.5%. Proctor and Hughes assure that this level allows for a good margin of safety.[2] The benefits of good forced-ventilation systems cannot be overstated. A number of types of poisonous gases can be controlled in this manner. Forced ventilation is also important because carbon dioxide is heavier than air and settles in the lowest areas of a building.

Carbon Monoxide — CO

Carbon monoxide is nothing more than the product of incomplete combustion. An internal combustion or diesel engine generate carbon monoxide. Cars and trucks, boilers, aerators and pumps are sources of this deadly gas. Carbon monoxide has only a threshold-limit-value (TLV) of 50 ppm[2]. No atmosphere, however oxygen rich, is safe if it has more than this amount of the toxin. This is because carbon monoxide aggressively competes with oxygen for the hemoglobin site; forming carboxyhemoglobin. Although this effect is reversible, the binding strength of this gas is 210 to 300 times greater than oxygen on the hemoglobin.[2,3] Even with the administration of 100% oxygen, the flush halftime of CO can be 80 min.[2] Any person suspected of having been exposed to this gas should be taken to a hospital for treatment. Do not rely on how the person feels. Call 911 and have the worker taken to the hospital. Lingering effects of carboxyhemoglobin production in the body can involve the heart, brain, and CNS. Symptoms can persist for days or even months after an incident. Lack of respect for this gas can result in permanent damage to a worker.

The symptoms of acute poisoning are dizziness, weakness, and mental confusion. The person may also have a peculiar cherry color of the nails and lips, due to the formation of carboxyhemoglobin.[2] Chronic exposure to this substance can result in headaches, irritability, and impaired judgment. Small amounts of this gas can aggravate preexisting heart problems. Smokers already carry a carboxyhemoglobin load of 2 to 10% in their bloodstream, so they are especially susceptible to this poison.

CHEMICAL HAZARDS — PULMONARY AGENTS

Pregnant women stand a good chance of damage to their unborn child from even low levels of CO. People with preexisting blood problems, such as sickle-cell anemia or other hemoglobin problems, can also be especially susceptible to the effects of carbon monoxide poisoning.[3]

One cannot overemphasize the importance of keeping combustion engines and exhaust systems in good working order and of having good ventilation. Winter, especially, is a bad time for CO accidents because most buildings are tightly closed to keep out the cold. But in all seasons one should be watchful in the workplace for sources of this gas, as it is colorless, odorless, and tasteless.

Methane — CH_4

The major source of methane at wastewater treatment plants is anaerobic digestion. It can also be found in poorly ventilated sewage pipes where the flow of water is sluggish, because of the anaerobic digestion of poorly aerated sewage. In addition to its asphyxiant hazard, methane presents an explosion danger. Its lower limit of flammability is about 6%, and the upper limit is about 15%.[4] In the range of about 6 to 10%, a methane/air mix can be positively explosive.[5] Just in the past year a sewage treatment plant in Ohio lost two men and had over two million dollars worth of damage due to a methane explosion. Methane and air mixes explode violently and easily. Even a static electricity spark can set off a methane explosion.

Because of the great combustibility of methane, certain safety rules must be stringently followed:

1. Positively no smoking or open flames should occur around anaerobic digesters, wet wells, or manholes.
2. Where methane flames are to be used in a lab setting, make sure all bunsen burners and equipment are in proper working order.
3. In the lab setting, make sure all valves are tightly shut off when work is completed.
4. Around digesters and other field sites, make sure all electrical equipment is of the explosion-proof variety.
5. Make sure that all valves, condensate traps, and flame traps around digester and Imhoff sites are clean and in good working order.
6. Set up regular inspection and maintenance schedules for all digester equipment.
7. When work must be done in an area, check the methane/air ratio. Use forced ventilation when necessary.[5]

Following such rules can prevent a tragedy from happening.

Methane is not toxic in and of itself. Since the brain is dependent on carbon dioxide to signal breathing, and not the lack of oxygen, the person may not show any symptoms before losing consciousness. This makes exposure to methane especially dangerous. If a worker is overcome with methane, the first aid is very similar to that for CO_2. Remove and resuscitate the person. Again, there should be a period of observation and follow-up by a qualified physician. Because methane is lighter than air, it is important that the air be circulated in a brisk way. Vigorous circulation and cross-ventilation ensures that no pockets of any gas form in either low or high areas.

Hydrogen Sulfide — H_2S

Hydrogen sulfide presents a number of problems. It is most often encountered in the field. Few, if any, reactions in the lab generate this gas. It is, however, a common and undesirable byproduct of anaerobic processes. This gas is unusual in that it is not only active on the lungs and mucus membranes, but it is also skin active. Since the gas is a byproduct of bacterial degradation of proteins, it is found in sluggish sewage lines, anaerobic digesters, wet wells, and lagoons. It is even important for research scuba divers to be extremely cautious. A hypolimnion of a badly polluted lake, or the lower regions of a meromictic lake, often contain large amounts of hydrogen sulfide. Certain water bodies have been known to have high enough concentrations of H_2S to present a problem to a commercial diver in a wet suit. The gas is also an explosion hazard when the hydrogen sulfide/air ratio is between 4.3 and 45.5%.[4] It is also corrosive. Major crown corrosion damage to sewage lines and the pitting of equipment have been attributed to this gas.

The toxicology of H_2S is complex. It is at least as toxic as cyanide. The 1977 threshold limit value-time weighted average (TLV-TWA) was set at 10.0 ppm.[2] However, sensitive individuals can complain of a rotten egg odor with as little as 45 to 130 ppb.[3] The detection of odor as a warning is unreliable, as the olfactory nerves soon become paralyzed.[2] You must, therefore, use a hydrogen sulfide detector before entering suspect areas. Since this gas is almost twice as heavy as air, it concentrates in dense pockets in low-lying areas. Even an open clarifier that has been drained for repairs can present a hydrogen sulfide hazard.

Acute pathology has a rapid onset. At levels of 400 to 2000 ppm, death can occur within seconds, due to paralysis of the breathing centers of the brain, located in the medulla.[2,3] There are other effects that must be considered at dosages between 50 to 250 ppm. This gas is a severe irritant, and mucus membranes can become damaged from as brief as 1 h of exposure. It is not unusual for the lungs to suffer severe edema that is, an excessive buildup of fluid.[3]

CHEMICAL HAZARDS — PULMONARY AGENTS

The chronic pathology that develops over a period of low-level exposure can also be serious. The brain, lungs, liver, and kidneys can be damaged. Corneal damage can also occur,[6] that is, the surface of the cornea can actually become blistered. With as little as 20 ppm, these effects can be felt.[3] Protective clothing must be worn in areas where the gas is above this level. A self-contained breathing apparatus is not enough, as hydrogen sulfide can be adsorbed through the skin. Again, a good ventilation system will not only save workers, but will also save equipment from corrosion, as hydrogen sulfide is highly corrosive. Be sure to check the ventilation equipment frequently for wear and tear, as the hydrogen sulfide could also erode some parts of the fans or blowers.

A rapid response to hydrogen sulfide injury is important. The rescuer should take care to use SCBA when entering a dangerous area. Rescuers have been known to die while attempting to remove a worker. These actions should be taken when dealing with a hydrogen sulfide injury:

1. Get the injured party out of the danger area.
2. Check for breathing and heartbeat.
3. Initiate resuscitation.
4. Gently flush the eyes and skin with water after the victim has been stabilized.
5. The worker should then be hospitalized for at least 72 h. Most of the lung edema and CNS problems will appear during this period.

Some workers will remain in a weakened state for months after the incident. Placement of the worker back on the job should reflect consideration for the worker's weakened state. Workers who have survived a hydrogen sulfide incident should be especially cautious when approaching areas suspected of having this gas. The author knows from experience that neurological effects can linger for a long time. This common byproduct of sewage treatment requires the utmost respect.

2.2 RESPIRATORY IRRITANTS AND OXIDANTS

Chlorine, chlorine dioxide, ozone, sulfur dioxide, ammonia, and various fuming acids are used for disinfection, chemical analysis, and pH control. Thus, the pollution worker can be exposed to these agents both in the field and in the lab. The most commonly encountered field agents are chlorine (and its related gases), sulfur dioxide, and ammonia. Common irritants in the lab are byproducts of some digestive processes or analyses. Gases encountered here are nitrous oxide (a common auxiliary gas used in atomic absorption), other oxides of nitrogen, and oxides of sulfur. In both the field and the lab, the worker should be familiar enough with basic chemistry to know which agents are compatible and which are not. A lot of these gases are generated as a result of oxidation-reduction reactions. When a gas must be used for a procedure, good ventilation practices must be observed.

Precautions for the Lab

There is no particular set of gases used in the lab. Most gas production is an incidental byproduct of some activity. There is, however, one exception. Nitrous oxide (N_2O) is a gas frequently used in atomic absorption for the analysis of transition metals. Some heavy metals require the hotter flame that a nitrous oxide/acetylene system produces. This gas is particularly destructive to the central nervous system (CNS). It is also a violent oxidant and even requires that a flame for atomic absorption work be lit in a specific way, using special attachments. Fuel-to-oxidant ratio recommendations specified by the manufacturer must be rigidly followed. Whenever possible, the chemist should avoid using this dangerous and unpredictable gas.

The toxicology of N_2O is particularly insidious, yet there are no standard TWA-TLV levels for this gas. Sweden sets an 8-h exposure at 100 ppm.[7] The U.S. National Institute of Occupational Safety recommends a level of no more than 25 ppm.[8] Dentists are required to have no greater than 50 ppm in the office atmosphere.[8] Certainly; after reading available literature on the toxicology of this gas, the author would recommend that all atomic absorption rooms be monitored on a frequent basis and that the level of N_2O be kept between 25 and 30 ppm. Not only should the hood of the atomic adsorption unit be turned on full; but forced ventilation via a strong fan should be used. If it is difficult to maintain the 25 to 30 ppm, the lab should consider the purchase of a scrubber. Be careful to check all attachments for leakage. Keep all gas lines in good repair and replace any worn lines. Nitrous oxide is a strong oxidant and could quite possibly attack some components of the instrument. The possibility of explosion can be reduced by following the manufacturer's recommendations for running the atomic absorption (AA) unit to the letter.

There is a real possibility that chronic damage can happen to the exposed chemist over a period of time. Nitrous oxide attacks several parts of the body. The gas effects are cumulative with each exposure. This gas is a deadly mutagen, its effects on the reproductive system will be covered in detail in another section. It is a particularly vicious oxidant, attacking the body's reserves of vitamin B_{12}. This vitamin is important in DNA and blood synthesis.[8] Even with B_{12} therapy, it may take several days for a lab technician to reverse the effects of B_{12} destruction. Nitrous oxide also attacks the CNS, kidneys, and liver.

Immune inhibition is also suspected.[8] It is good for the chemist to be aware of the danger and take steps to avoid trouble.

Always, in approaching the day's work the chemist should be mindful of the possible safety problems inherent in a particular analysis. All new samples should be prepared under the gas hood, since many wastes contain cyanides or other substances capable of generating gases upon the addition of acid or base. The digestions of samples for metals should be handled with care. It is not at all uncommon for soil or sludge samples to evolve nitrogen oxides and chlorine during an acid digestion. Nitric acid slowly breaks down on its own in the presence of light. The byproducts are nitric oxide and nitrogen dioxide.[2] Fumes of both nitric and hydrochloric acid can erode tooth matter over a period of chronic exposure. Chlorine and nitrogen oxides are both strong pulmonary irritants capable of causing severe edema or fluid in the lungs several hours after exposure. None of these gases should have a TLV-TWA of over 2 to 5 ppm. It cannot be overemphasized that the treatment of injuries resulting from exposure to these gases requires the utmost care. Symptoms are often delayed, so hospital observation of the victim for 24 to 48 h after the injury is important.

Dangers in the Field

Chlorine and Chlorine Dioxide — Cl_2, ClO_2

Chlorine and its related compounds are the most commonly used of all disinfectants. Chlorine gas accidents are some of the most common occurrences in water and wastewater treatment plants. Both gases are deadly. The TWA-TLV of chlorine is 1 ppm.[2] The value for chlorine dioxide is only 0.1 ppm.[2] Chlorine is shipped in 100-lb cylinders up to railroad tank car-size, in accordance to a particular plant's needs. Sometimes calcium hypochlorite or sodium hypochlorite is used in package plants, rather than a gas. These salts, however, can generate chlorine gas or even be the cause of explosions, under the right conditions. Chlorine dioxide is commonly manufactured on site by the reaction of chlorine gas with sodium chlorite:

$$Cl_2 + 2NaClO_2 = 2ClO_2 + 2NaCl^9$$

The gas is highly explosive and is used immediately after generation. This agent, despite its safety problems, is preferred for some taste, odor, and color problems. Both chlorine and chlorine dioxide have serious handling problems and require stringent storage and safety policies.

Injuries resulting from chlorination exposure are serious. Rescuers have also died in attempting to remove an injured party from a leakage area. No disinfection area should be without a complete and working

14 CHEMICAL HAZARDS AT WATER TREATMENT PLANTS

A

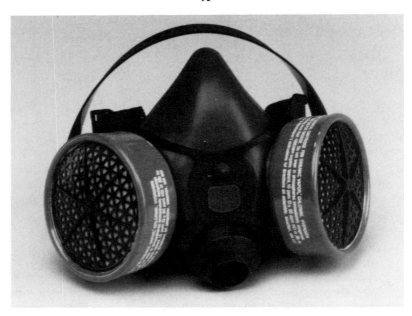

B

Figure 2.2 Cartridge Respirators for use only in O_2 adequate environments. Make sure to use the right filter for the right gas.

CHEMICAL HAZARDS — PULMONARY AGENTS

self-contained breathing apparatus. The apparatus should be kept in an easily accessible area outside the chlorine room. Chlorine cartridge respirators are only good for protection from light contamination. (see Figure 2.2). They are a poor choice for really serious spills or leaks. If an accident occurs, the first order of business should be to protect oneself with a SCBA and then remove the victim from the danger area. Speed is of the utmost importance. As little as 19 ppm of chlorine dioxide has resulted in the death of a worker.[2] Check for breathing and heartbeat. If need be, initiate CPR or artificial respiration. Once the person is stable, get him to a hospital. No worker that has been exposed to a chlorine agent should be allowed to go home on his or her own, no matter how well he or she feels. An observation period of at least 72 h should be observed. Pulmonary edema and collapse has occurred as late as 2 or 3 d after an incident. Also, a type of traumatic pneumonia has occurred after a serious chlorine injury.[3] Have the ambulance personnel drag your worker to the hospital if you have to. He'll thank you for it later.

At no time should the employer take a casual attitude towards good industrial hygiene. The chronic effects of both gases, over a period of exposure, can be devastating. It is interesting to note that the U.S.S.R. requires a TLV that is a third of the United States level.[3] When one looks at the effects of long-term chronic exposure, one can see why levels of both gases should be kept far below 1 ppm. Symptoms appear very slowly over a period of years. People age prematurely, teeth begin to "rot" out due to hydrochloric acid forming in the mouth, and some people become predisposed to tuberculosis and emphysema.[3] A nonallergic type of asthma may also be brought on by low level chlorine and chlorine dioxide exposure. There are certain rules, set by the Chlorine Institute, for the handling and storage of these gases and other chlorine compounds. These should be followed to the letter: NO short cuts!

1. Chlorine facilities should be kept in a completely separate aboveground building that has it own separate drainage and ventilation system, since these heavier-than-air gases will collect in the lowest point of a plant.
2. View windows, which allow an operator to check the building without entry, should be installed.
3. Forced ventilation should be tripped automatically when an operator enters the chlorine facility.
4. Forced ventilation should provide at least 15 air volumes of circulation per hour.
5. Any light and fan switches should be placed outside the building so the operator can blow the area out before entry.
6. **NO** chlorine compound should be stored anywhere near organic matter of any kind. Chlorine-organic mixes are explosive.

7. The operator should, when connecting new tanks, carry a light solution of ammonium hydroxide with him at all times, for leak testing. When a light solution of NH_4OH is held near a leaky valve or connection, a white smoke of NH_4Cl will form as ammonia vapor is wafted close to the connection.
8. The area itself should have an alarm system sensitive to 1 ppm of chlorine gas.
9. All breathing equipment should be in a ready state. One SCBA should be in a highly visible area outside the building, and another in the building. These units should be frequently checked and kept clean and in good repair.
10. Training should be provided for all operators handling chlorine. Training should include chemistry and safety.[10]

These precautions are meant to keep tragedy and disability from happening. They are a cheap investment when one considers the loss of earning power that occurs when a worker is injured.

Sulfur Dioxide — SO_2

It is common in some establishments to have a dechlorination step after disinfection. The reaction is as follows:

$$SO_2 + H_2O = H_2SO_3$$

$$H_2SO_3 + HOCl = H_2SO_4 + HCl$$[10]

This gas is a strong oxidant and highly irritating to the lungs. The gas is usually supplied in ton containers or rail tank cars. Because of the danger inherent in gas storage and handling, some plants opt for using other agents for dechlorination, such as sodium bisulfite, sodium sulfite, and sodium metabisulfate. These agents are not completely fool proof, however. Under certain conditions these salts can generate unwanted sulfur dioxide. They can react rather violently with certain acids. These salts are susceptible to degradation and should be stored in cool, dry areas separate from other agents.

Sulfur dioxide's TLV is set at 5 ppm, but compromised individuals, such as asthmatics and smokers, could be harmed with as little as 1 ppm, especially when engaged in heavy labor.[2,11] Because so little sulfur dioxide can affect the respiratory system, the rules for handling it are much the same as for chlorine. Because sulfur dioxide forms a rather strong acid upon reaction with moisture, it can be a factor in corrosion. One should take steps to control humidity as well as ventilation in an area where this gas is used. Injuries due to sulfur dioxide must be attended immediately. As little as 10 to 50 ppm for 5 to 15 min can cause constriction of the bronchial tubes, nose bleed, injury to the mucus membranes, and a serious edema of the lungs.[2] Again, this is a gas that requires a 72-h observation period after the initial first aid. Since as little as 20 ppm can cause respiratory paralysis, a rapid response by a person experienced in CPR is important. There is another complication to injury with this gas; about 10 to 20% of the population can be expected to show an allergy to sulfur dioxide.[2] As little as 0.75 ppm can cause changes in the cilia of the upper respiratory system in some people.[12] Repeated exposure to low levels of sulfur dioxide can result in asthma, chronic bronchitis, and emphysema. Since some people can be just as allergic to sulfur dioxide as they are to pollen, an exposure to the gas could force them to stop handling the gas even in ordinary working conditions. Employers should be on the lookout for a history of sulfur compound sensitivity in their employees and take care to place workers in tasks safest for them. With a little caution, handling sulfur agents need not be dangerous.

Ammonia — NH_3

Ammonia is rarely generated in wastewater in amounts that can be an atmospheric hazard. However, pure ammonia gas is used in certain drinking water applications. The ammonia is combined with chlorine to produce chloramines. Both mono- and dichloramines persist in the distribution system longer than does free chlorine. Hence, they have a more effective residual in a large distribution system. The balance is very fine, however. If too much ammonia is applied, then trichloramine forms. This is a vile smelling and tasting form of combined chlorine, which has absolutely no disinfection power. Reactions are as follows:

$$NH_3 + HOCl = NH_2Cl + H_2O$$

$$NH_2Cl + HOCl = NHCl_2 + H_2O$$

$$NHCl_2 + HOCl = NCl_3 + H_2O\;[9]$$

The TLV of ammonia is 25 ppm.[2] Although roughly 250 ppm can be highly irritating, it is not as dangerous as some of the gases already covered. This is because the body has some limited ability to neutralize ammonia via combination with the carbon dioxide and moisture of the lungs.

$$CO_2 + H_2O = H_2CO_3$$

$$NH_3 + H_2O = NH_4OH$$

$$2NH_4OH + H_2CO_3 = (NH_4)_2CO_3 + 2H_2O$$

However, one must not become complacent. A concentration of about 2500 to 6500 ppm can cause bloody edema and permanent injury to the lungs.[2] Ammonia is also violently reactive with certain metals and organic matter. Care must be taken to see that ammonia is stored in an absolutely dry condition in the proper holding tanks. In certain rare instances, ammonia has been known to form explosive mixes with air. For this reason some plants use ammonium hydroxide or ammonium sulfate.[9] Again, forced ventilation in areas where ammonia is stored and used is necessary.

First aid for ammonia injuries is dictated by the nature of the exposure. If the hydroxide is spilled on the skin, the injured area must be thoroughly flushed with copious amounts of water. The area should then be wrapped in clean, wet cloth or gauze. The person should then be taken to the hospital for burn treatment. A quick response and thorough cleaning of an ammonium hydroxide burn can mean the difference between permanent scarring and complete healing. Eye injuries are particularly dangerous. After the eye is forcibly flushed for 15 min with a gentle but continuous stream of water, the person should be taken to the hospital. Make sure the lids of the eyes are held open for irrigation, to assure complete cleansing. If only one eye is affected, be extremely careful to direct the irrigation AWAY from the uninjured eye. Follow up care by a trained ophthalmologist is important, as, very often, corneal scarring can occur with this kind of injury. To avoid such a serious eye injury in the first place, employees should be warned never to wear contact lenses on the job, as contacts will hold the poisonous agent against the eye.[4] Contacts may not be able to be removed by the primary rescuer. This may require a doctor's skill. Make sure all workers wear protective goggles and gloves when working with ammonium compounds. If it is a gas exposure, remove the injured party from the area and resuscitate as needed. As with all pulmonary irritants and oxidants, a 72-h observation period in a hospital is recommended. All of the gases covered thus far are capable of causing a delayed and severe edema.

Ozone — O_3

Ozone has been used for years in Europe as a disinfecting agent. It has only been in the last 15 or 20 years that the United States has looked at ozone for disinfection purposes. Some water quality personnel prefer this agent over chlorine, because the oxidation of taste and odor agents is more complete than with chlorine. Also, in areas where raw drinking waters are high in natural organic substances, ozone may be the gas of choice, as trihalomethanes (THM) are not formed. The gas is generated by passing a very strong electrical current through dry, clean compressed air. The dryness and cleanness of the air is important, as ozone generators have been known to explode if these conditions are not met.

Of all the gases covered thus far, ozone is the most toxic. The current TLV is set at 0.1 ppm.[2] There are individuals that show changes in vision with as little as 0.05 ppm, due to possible effects on the eye muscles.[3] With as little as 10 to 12 ppm, exposure to ozone could be rapidly lethal.[3] This gas is a severe oxidant and can attack cellular components directly. Although humans show some ability to develop a certain level of tolerance to chronic exposures at 0.1 to 0.18 ppm, the body of evidence shows that long-range damage can occur to the body over a long period of time.[13] These effects are

1. Chronic pulmonary disease — Ozone directly attacks the cilia, lung epithelium, and lymphocytes.[14] With high levels, pulmonary edema can occur. Over years of low level exposure, one can develop emphysema and fibrosis. The ability to fight off upper respiratory infections could also be hampered.
2. Premature aging — Experimentation with animals show that premature aging occurs. Symptoms are dulling of the cornea, stiffening of lung tissue, and depletion of body fat.[3] Although there is no direct evidence of this in humans, it is good for the worker to take every possible precaution to limit his exposure to this gas.
3. Growth of lung tumors[3] — Human evidence in this area is sketchy at best, but the fact that ozone can interfere with the enzymes that aid in the metabolism of RNA should be a warning to workers of the possibility of ozone-induced cancer.

Because of these effects, any plant considering ozonation should carefully consider the safety features of the ozone generator. The room in which the generator is kept should have a detector and an alarm system set at 0.1 ppm. Forced ventilation should be provided in the ozonation area. A self-contained breathing apparatus should be kept in a prominent, visible location. The breathing gear should be checked out and kept in good repair. Air fed to the ozone generator needs to be passed through a filter and drying system. Provisions should be made to keep an ozone generator from overheating. Some systems use pure oxygen as the feed gas. If this is the case, then provisions must be made for the safe handling of pure oxygen. Absolutely no open flames, sparks, or smoking should be allowed in the oxygen storage area.

A short summary of the pulmonary irritants is provided at the end of the chapter (Table 2.1). This table of chemistry and first aid is by no means complete. When setting up a plant for the handling of any gas, one should work closely with the manufacturer of the equipment and gas being purchased.

Table 2.1. Gas Chemical Properties and Safe Handling Recommendations

Compound	Source	Storage	Precautions	First Aid
Carbon dioxide (CO_2)	Biological activity	Generate on site if used in process	Ventilation	Remove and resuscitate
Carbon monoxide (CO)	Incomplete combustion	N/A	Ventilation; keep motors in good repair	Remove and resuscitate
Methane (CH_4)	Anaerobic biological activity	N/A	Ventilation; avoid open flames and sparks	Remove and resuscitate
Hydrogen Sulfide (H_2S)	Anaerobic biological activity in septic processes	N/A	Ventilation; test for gas before entering work area	Remove and resuscitate; flush eyes gently with water
Nitrous oxide N_2O	Commercial gas for AA	Tank in cool room	Ventilation; follow AA recommendations	Remove and resuscitate; vitamin B_{12} 3 mcg oral
Chlorine and related compounds (Cl_2, ClO_2, OCL)	Commercial chemical	Tanks; resistant drum; separate cool, dry room	Ventilation; not compatable with organics or other oxidants	Remove and resuscitate; flush burns with water
Sulfur dioxide and related compounds (SO_2)	Commercial chemical	Tanks; resistant drum; cool, dry room	Ventilation; not compatable with organics or other oxidants	Remove and resuscitate; flush burns with water

Table 2.1. (continued)

Compound	Source	Storage	Precautions	First Aid
Ammonia and related compounds (NH_3)	Commerical chemical	Tanks; resistant drum; dry room	Ventilation; not compatible with organics or other oxidants	Remove and resuscitate; flush burns with water
Ozone (O_3)	Generated on site	N/A	Ventilation; positively no moisture or organics; no sparks or flames	Remove and resuscitate; monitor for asthma and emphysema

REFERENCES

1. Berne, R., and M.N. Levy. 1988. *Physiology*. C.V. Mosby, St. Louis. p. 627.
2. Proctor, N.H., and J.P. Hughes. 1978. *Chemical Hazards of the Workplace*. J.B. Lippencott, Philadelphia.
3. Waldbott, G.L. 1978. *Health Effects of Environmental Pollutants*. C. V. Mosby, St. Louis.
4. Steere, N.V., Ed. 1971. *CRC Handbook of Laboratory Safety*. 2nd ed. CRC Press, Boca Raton, FL.
5. New York State Department of Health. *Manual of Instruction for Sewage Treatment Plant Operators*. Health Education Service, Albany, NY.
6. Layton, D.W., and R.T. Cederwall. 1986. Assessing and managing the risks of accidental releases of hazardous gas: a case study of natural gas wells contaminated with hydrogen sulfide. *Environ. Int.* 12:519–532.
7. Munley, Railton, Gray, and Carter. 1986. Exposure of midwives to nitrous oxide in four hospitals. *Br. Med. J.* 293(25):1063–1064.
8. Schumann, D. 1990. Nitrous oxide anaesthesia: risks to health personnel. *Int. Nursing Rev.* 37(1):214–217.
9. Weber, W., Jr. 1972. *Physicochemical Processes for Water Quality Control*. John Wiley & Sons, New York.
10. Joint Committee of the Water Pollution Control Federation and American Society of Civil Engineers. 1982. *Wastewater Treatment Plant Design*. Lancaster Press Inc., Lancaster, PA.
11. Linn, W.S., O.A. Ficher, D.A. Shamoo, C.E. Spier, L.E. Valencia, U.T. Anzar, and J.D. Hackney. 1985. Controlled exposures of volunteers with chronic obstructive pulmonary disease to sulfur dioxide. *Environ. Res.* 37:445–451.

12. Carson, J.L., A.M. Collier, Shih-Chin Hu, C.A. Smith, and P. Stewart. The appearance of compound cilia in the nasal mucosa of normal human subjects following acute, in vivo exposure to sulfur dioxide. *Environ. Res.* 42:155–165.
13. Tilton, B.E. 1989. Health effect of tropospheric ozone. *Environ. Sci. Technol.* 23(3):257.
14. Hark, E.D., Jr. 1981. Human pulmonary adaptation to ozone. In Research Planning Workshop on Health Effects of Oxidents. EPA-600/9-81-001. U.S. EPA, Washington, D.C.

3

Chemical Hazards — Heavy Metals

3.1 SOURCES OF CONTACT

In both the field and lab there is ample opportunity for the worker to be exposed to heavy metals. Some of these metals, such as the coagulants, can irritate and burn the skin if handled incorrectly. Others, such as cadmium or mercury, can be quite deadly even in small amounts. Therefore, it is important that the employer be aware of the potential for heavy-metal exposure in the place of work. Then steps can be taken to avoid contamination and injury.

There are three major areas of exposure to heavy metals in the water and wastewater treatment plant. One area that presents a potential for especially heavy exposure is the lab. Injuries can also result from the use of acidic coagulants such as alum or ferric chloride. Lastly, welder's or metal-fume fever can result from the careless handling of soldering or welding equipment in the repair shop of the plant.

The lab offers the highest likelihood of direct exposure. The potential areas of greatest exposure in the lab are atomic absorption analysis (AA), total Kjeldahl nitrogen analysis (TKN), and jar testing. In atomic absorption one can get a sickness very similar to welder's fever, if the work area is not properly ventilated. Kjeldahl nitrogen, surprisingly, is a source of mercury, due to the use of mercuric oxide as a catalyst. Jar testing deals with such acidic salts as ferric chloride and alum. Most hazards can be avoided by following industrial hygiene procedures.

Ventilation

When setting up the lab for analysis, one should provide for adequate ventilation. The AA unit should be provided with its own hood. The hood should be positioned in such a way as to have the greatest draw directly over the flame. All connections on gas lines and regulators should be leak tested with an approved bubbling agent. The nebulizer, head connections, and other parts should be kept in good working order. Make sure that the correct fuel-to-oxidant ratios for analysis are used. This will insure the correct flame velocity so that the gases don't blow back on the technician. To check to see if the hood is drawing correctly, use a smoke taper or a velocity meter. The smoke from the taper should be drawn straight up, or the face velocity at the hood should read between 75 to 100 fpm.[1] Another source of fume protection is the gas trap. This is usually a piece of looped tygon tubing connected from the nebulizer exhaust to a large teflon reservoir half full of water. This gas trap prevents the leakage of mixed gases to the atmosphere and also prevents blow backs. The gas trap also collects the excess solution from the aspiration of samples. There are special precautions that one must follow in handling mercury compounds. Because of this, mercury hazards will now be covered in a separate section.

Mercury — Hg

"Twinkle, twinkle, little bat! How I wonder what you're at .... Like a tea-tray in the sky."[2] This is not an attempt at entertainment, but a quote from the mad hatter in Lewis Carroll's *Alice's Adventures in Wonderland*. The portrayal of the mad hatter is entirely accurate. Lewis Carroll's fairy tale may be the first medically correct description of the toxic effects of mercury. The disease portrayed is the syndrome of mercury psychosis. Victorian hat makers used both elemental mercury and chlorides of mercury for their hat-making processes. It was not uncommon for them to start going quietly crazy over a period of years.

If mercury is in the elemental state, its TLV is 0.05 mg/m^3, or about 0.05 ppb. If it is in the organic form (R-Hg-X), the TLV is less than 0.001 ppm.[3] The amounts are small because mercury not only attacks the central nervous system, but also persists in the body for years. Its biological half-life is about 70 d. The problem with mercury is not the possibility of immediate injury. The real problem is that the effects of overexposure to mercury accumulate slowly over a period of months and years. The worker may slowly develop tremors of the hands, psychological problems, weight loss, and kidney trouble. Mercury may accumulate in a mother's fetus without her feeling the effects, since mercury preferentially crosses the placenta, that is, mercury will accumulate in the placenta and fetus before lodging in the mother's tissue.[4] Mercury easily crosses the skin barrier. The metal readily vaporizes at room temperature, and has no odor or taste. Because mercury is preferentially absorbed by nerve, fat, and kidney tissue, a blood test may not detect mercury exposure right away.[4] Tissue biopsies may be needed to confirm exposure. There is no first aid for mercuric exposure except to remove the person from the area of danger. The worker's primary physician would probably start treatment with a chelating agent called BAL or dimercoprol. It takes months to heal from the effects of mercury and the side effects of the treatment itself. Damage may be irreversible.

Because of the very serious nature of mercuric poisoning, one must follow stringent practices of lab safety. Employers should require chemists working with mercury to wear a monitor badge of palladium chloride. The author would also recommend testing hair, blood, and urine at least once a year for mercury. Employers should be on the lookout for changes in employee behavior. Behavior that cannot be attributed to alcoholism, family problems, or other personal problems should be a flag for possible chemical-exposure problems.

Sources

Two major sources of mercuric exposure are fumes from TKN analysis and cold-vapor analysis for mercury. It is vitally important that the lab be set up to avoid the escape of vapors into the workplace. Use gloves when working with mercury salts. Run all mercuric digestions under the hood. A cold-vapor setup for AA analysis is a completely closed system. The prepared sample is dosed with stannous chloride, and a bubbler is attached to the bottle. The vapor is swept into a quartz optic chamber and read by the AA unit. A release valve directs the spent mercury vapor to a carbon trap (Figure 3.1), and the next sample is run. Even when using a carbon trap, one should make sure the fume hood of the atomic absorption unit is running at its highest setting: about 100 to 150 fpm.[1] Remember, the carbon trap will become depleted after several uses. Replenish the carbon in the trap every few weeks. Discard the old carbon filling in an airtight canister. This waste canister, when full, is a hazardous waste and should be disposed of properly. When working with TKN digestions, one should prepare the catalyzed digestion reagent under the hood. The digestion apparatus itself should be completely enclosed by an acid-resistant gas hood, or provisions should be made to vent the sulfur dioxide and mercury fumes to a scrubber. See Figures 3.2 and 3.3 for the proper provisions for mercury fume control.

Mercury sickness or the appearance of mercury in the tissues of a worker cannot be tolerated. The appearance of any mercury in the general lab atmosphere is a signal that fume hoods are not operating properly. Labs should be designed to protect the weakest individual from mercury poisoning.

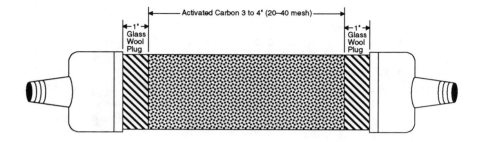

Figure 3.1 Carbon trap (or mercury trap). This trap is attached to the outlet hose under the fume hood of the AA. If use is heavy, change the trap weekly. See that the waste is put in an airtight container and properly disposed. (Based on Perkin-Elmer 4000 ©1980.)

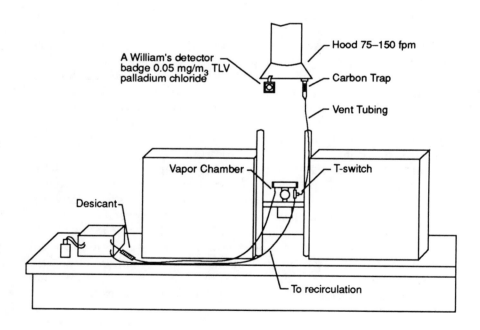

Figure 3.2 A safe mercury vapor setup. Note: not only should a 0.05 mg/m^3 detector be used at the hood, but the technician should also wear a badge on his person. (Based on Perkin-Elmer 4000 ©1980.)

CHEMICAL HAZARDS — HEAVY METALS

Other Dangers of the Lab and Shop: Welder's Fever

An atomic absorption flame or a welding unit are sources for small particles of metal oxides. Certain metals, such as iron or magnesium, may not be toxic themselves. But in an oxidized condition, only a few ppm of metal fumes can cause a debilitating condition called metal-fume, or welder's, fever. Symptoms appear a few hours after exposure and mimic the flu. An episode of illness can be as brief as 12 h or last as long as 48 h. The worker experiences dry cough, high fever, chills, nausea, and aches and pains. Complete prostration is not uncommon. It is a very uncomfortable, if not serious, illness. This disease is self limiting, but bed rest and plenty of fluids are recommended. Needless to say, providing proper ventilation around metal-repair areas and lab units can cut down the number of work hours lost to sickness. No welding should be done without a vigorous rate of cross-ventilation. Electricians in the plant should be careful in handling solder, as this is also a source of lead.

Other Heavy Metals in the Lab

There are other heavy metals that are significant to the lab and plant worker. Some are present in polluted samples, and some are used as reagents or cleaning agents. The most important metals in this respect are lead, manganese, chromium, cadmium, and beryllium. These metals are toxic in small amounts and deserve the respect of the worker.

Lead — Pb

Lead is a serious problem for analysts, operators, and handymen. Although lead has been removed from most gasoline, it is still a component of many polluted samples and is present as an impurity in many metals. All sources of lead should not exceed a combined TLV of 0.15 mg/m^3.[3] This metal, like mercury, is also preferentially adsorbed by a fetus. This metal, along with other agents, will receive special consideration in a section about reproductive hazards. For now, the dangers to the primary handler will be covered.

Lead poisoning, or plumbism, as its called, has been known for over 2000 years. There are varied sources of lead for the average citizen, such as pottery, paints, and certain pesticides. For the plant operator or chemist, sources of lead come from soldering activities, welding, and atomic absorption work. Like mercury, lead can slowly accumulate in the body over a period of time. Its biological half-life can be several years. Indeed, finds from digs at Gloucestershire show that the ancient bones of Roman aristocrats still contain significant amounts of lead.[5]

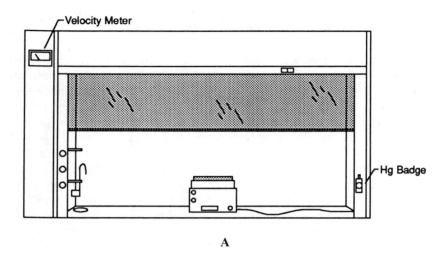

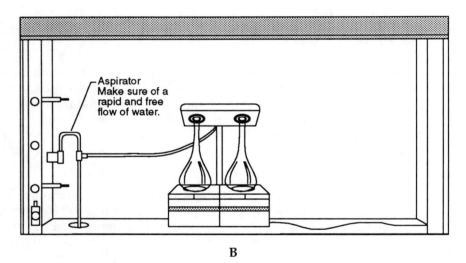

Figure 3.3 (A) A safe setup for a micro TKN Digestion ("100 ml). Note the William's badge for Hg fume detection. Make sure face velocity of the hood is at least 75 fpm. (B) Close-Up of a macrodigestor (100 to 800 ml). Note: a water aspirator carries off all fumes. Still, keep face velocity of the hood 75 fpm or more.

Lead accumulates in the bones and teeth. It will also reside in the kidneys and CNS. Because of its marked preference for bone tissue, it is difficult to treat with chelating agents. (A chelating agent is a chemical that combines with a heavy metal so that it can be excreted from the body.) The bones can serve as a reservoir for lead long after the initial poisoning. Lead can become mobile during periods of stress and growth or if the diet is lacking in vitamin D and calcium. As the body takes reserve calcium from the bone, it frees the lead into the blood.

The diagnosis of lead poisoning can be difficult. It mimics a baffling variety of diseases. Symptoms include moodiness and violent outbursts of temper, gout[6], digestive disturbances, anemia, and a lack of coordination. Any employee showing these symptoms should be screened for lead poisoning. Since man can only tolerate a daily burden of 1 or 2 mg of lead per day from all sources, lab personal should be very careful in handling lead compounds.[7] Here again, a sound ventilation system can be a successful first line of defense.

Manganese — Mn

This is a common trace metal in well water. Therefore, the chemist is frequently requested to run analyses for Mn. This is the first metal mentioned that is indeed also a trace nutrient as well as a poison. Dosage is everything. We certainly get enough Mn (about 3 to 9 mg/d)[7] in the foods we eat. Whole grains, eggs, and green vegetables are common sources of Mn. So we certainly do not need it in the air or water around us. Normal body burden for Mn is about 10 to 20 mg per body, and the organs that retain the most Mn are the bones, liver, pancreas, and pituitary gland.[7] Vapor TLV for Mn is set at 5 mg/m^3.[3] In water it is recommended that Mn not exceed 0.05 mg/l, not only because of its toxicity, but because it gives well water a very unpleasant, sweet, metallic taste. Water quality workers can be exposed to Mn via the handling of potassium permanganate used in certain oxidation processes and in heavy metal analysis. Potassium permanganate ($KMnO_4$) is shipped in a granular form and is an oxidant used for taste, odor, iron, and manganese removal in water treatment.[8] The compound is a strong oxidant. Aside from its primary toxicity, it can also burn and discolor the skin.

When handling $KMnO_4$ either in bulk or as a reagent, one should wear gloves. In handling large amounts for water treatment, it might be advisable to wear a protective face mask as protection against the dust. Goggles are also in order, since any oxidant can injure the eyes. If the solution or powder gets on the skin, one should flush the affected area with water for 15 min. The author has been very successful in removing permanganate stains, using ascorbic acid, i.e., vitamin C. After flushing the skin thoroughly, one simply applies a paste of vitamin C to the brown stain for 1 to 2 min. This reduces the Mn^{+7} to Mn^{+2}. After the stain is removed, flush with water again for another 5 min to remove all traces of Mn from the skin. If potassium permanganate is splashed in the eyes, flush immediately for 15 min with plain water. DO NOT attempt to neutralize any eye injury with an antidote. Get the injured worker to a trained ophthalmologist immediately!

Because this metal in its +7 state is a violent oxidant, proper storage is important. Potassium premanganate can be a fire hazard. It is usually shipped in 110- to 600-lb chemically inert drums, for water treatment purposes. It needs to be stored in a cool, dry place. Potassium permanganate should never be stored around other organic matter. Never store this chemical near reducing agents. Proper handling and storage can prevent problems.

Chronic intoxication with this metal involves the central nervous system and mimics the symptoms of Parkinson's disease.[3] There can also be a pattern of compulsive obsessive behavior with manganese poisoning.[4] If a worker shows any of these symptoms, he should be checked out by a qualified environmental physician. It is easy to prevent such problems simply by following good laboratory hygiene. Cleanliness and good ventilation are the first line of defense in safety.

Cadmium — Cd

The only way any water quality worker would be exposed to cadmium is during chemical analysis. Industrial samples are analyzed for Cd, and Cd is used as a reducing agent for nitrate analysis. It is fortunate that exposure to this metal is limited. It is an extremely toxic agent. The TLV for fumes and dust is no more than 0.05 mg/m^3.[3] As little as 0.5 to 2.5 mg/m^3 can cause a form of pneumonia. As little as 9.0 mg/m^3 can kill in a matter of hours. Because the amounts of Cd handled are small, it is unlikely that an acute exposure could happen. Chronic exposure is another matter, however. Cadmium attacks the kidneys, bones, and lungs. The famous "itai-itai" or "ouch-ouch" bone disease of the Japanese in Toyama Inlet was a result of chronic exposure to Cd wastes.[4] Itai-itai means "ouch! ouch!" in Japanese. The name is descriptive of the intense pain felt in the bones of the sick person. The disease further progresses to a bone weakness that results in multiple stress fractures in the back and legs. The biological half-life of this element is not known. But it was found that the average American, upon death, has a body burden of 30 mg.[4] This indicates that cadmium may reside in the body for years after ingestion.

There is but one remedy for Cd hazards. That is stringent lab hygiene. When preparing Cd reduction columns for nitrate analysis, one should work under the hood and use gloves. Make sure all hoods used with AA work are in good working order. Always completely clean work areas where Cd was handled. Wash hands thoroughly after working with Cd. There is simply no excuse for a worker to suffer Cd intoxication. A few simple precautions can avoid big trouble concerning this substance.

Beryllium — Be

Beryllium is not a problem except when a chemist analyzes a sample for this element. However, this element is so toxic that some chemists refuse to work with suspected samples without a self-contained glove box. Its TLV is only 0.002 mg/m^3.[3] This substance attacks just about every organ in the body. If splashed on the skin the compounds can cause suppurating ulcers. If inhaled, a severe form of pneumonia follows. An acute episode can last up to 3 months, and very often the acute disease goes into a lethal form called granulomatous inflammation.[4] This is a form of clustered lung tumors that eventually grow to crowd out the useful lung tissue. It may only take 20 to 60 μg per man per day for a few weeks of chronic exposure to the dust and fumes to cause this horrible sickness.[4] If your lab does not have clean hoods or closed glove-box areas for analysis of samples suspected of containing Be, it may be better to contract this type of work out to a qualified lab. This is not a metal the amateur analyst can safely handle.

If any environmental worker is suspected of being exposed to this metal at a waste site, flush the exposed skin with copious amounts of water. Every trace of the chemical must be washed off. Destroy any clothing contaminated with beryllium. The person should be observed at a hospital for at least 72 h for lung disease. This disease may require steroid therapy and demands the skills of a specially trained environmental physician. Teach your workers to respect beryllium.

Chromium — Cr

In the wastewater and water setting, chromium is used as an oxidant in the COD (chemical oxygen demand) test and as a cleaning agent. It is also one of the most commonly found metals in industrial wastes. This is also an odd metal in that it is also a micronutrient, but, the dosage is extremely small. In a diet including clams, corn oil, and whole grain cereals, man may take in 80 to 100 μg/d.[7] Most high potency vitamins contain only 25 μg per dose. Although chromium is needed for normal sugar digestion, a human should have no more than a total of 100 μg/d from ALL sources. The OSHA standard of safety for chromium fumes is 1 mg/m^3.[3] However, some literature indicates that this standard should be reduced to 25 μg/m^3 or less.[4] In certain sensitive individuals, a Cr dose as low as 2.5 μg/m^3 can cause changes in the eyes' adaptation to darkness.[4]

Chromium exists in three oxidation states: the neutral (0) state, which is found in plating and cheap jewelry; the $^{+3}$ state of foundry dust and spent oxidants; and the $^{+6}$ state of oxidizing and plating agents. Some

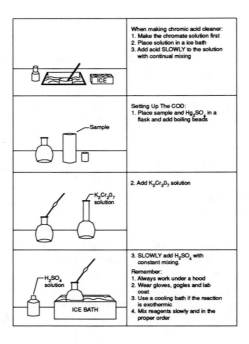

Figure 3.4 Safety with cromate salts.

people are so sensitive to Cr that even the 0 state can cause contact dermatitis. It is the Cr^{+6} state that is the most toxic, however. Unfortunately, it is also this state, in $K_2Cr_2O_7$ and K_2CrO_4, that is most frequently encountered by water quality personnel.

As stated earlier, the $+6$ state of chromium is the most toxic. Because Cr in this state is a strong oxidant, it can burn severely. A H_2SO_4 and potassium dichromate ($K_2Cr_2O_7$) mix, called chromic acid, is often used as a cleaning agent. Mixing sulfuric acid with chromate salts can be explosively exothermic, that is, the reaction rapidly generates large amounts of heat. Before preparing digestive reagents for COD or cleaning agents, one should wear a full lab smock, goggles and protective gloves. If one uses Chromerge® (a glass cleaner) with their acid, one can add the Chromerge® to the acid with constant mixing. If one makes one's own saturated potassium dichromate solution, one should add the acid TO this solution slowly, with continual mixing. Since heat evolution is quite rapid and violent, do this in a cooling bath under a hood. TAKE YOUR TIME. Don't rush when working with chromates or any acid. Make sure chromate cleaning solution is kept in its own special container away from other chemicals. When setting up a COD test (Figure 3.4), the reagents are usually added in this order: mercuric sulfate, sample, potassium dichromate solution, and then the sulfuric acid SLOWLY and with cooling. Mistakes are usually made when the chemist is daydreaming or in a hurry. When storing chromate compounds, make sure they are tightly sealed in a cool, dry place. DO NOT store chromate salts near any reducing agents, organics, or paper.

Sometimes, despite all of these precautions, accidents do happen. If a chemist splashes a chromate mixture in the eyes, the eyes must be completely flushed out with a constant stream of water for 20 min. The eye is a delicate organ. Once the eye is flushed, call 911 for an ambulance. Never attempt to put any agent other than clean water in the eye. Leave neutralization to the ophthalmologist. If the skin is contaminated, flush it with water for 20 min. Next apply a paste of sodium bicarbonate for 1 min. Then flush again with water for an additional 5 min. Then apply a 5% solution of ascorbic acid to reduce any traces of Cr VI, for about 1 min. Then flush another 5 minutes with clean water. Wrap the affected area in a clean, wet cloth and get the person to a hospital by ambulance. Discard any clothing exposed to the chromate spill. Even if it appears that injury has been prevented, it is important that a qualified doctor examine the affected area. Chromate salts are more than suspected of causing cancers as well as contact dermatitis.[9]

When working with chromate powders it is important that the work be done under a hood with protection. The respiration of chromate salts and dust has resulted in perforated septums, nasal polyps, fibrous tumors in the lung, and ulcers on the skin, and sometimes in the mouth.[4] Devotion to safety and cleanliness in the lab can prevent problems with chromates. Every chemist who works with these salts should have training in handling these compounds.

3.2 SPECIAL DANGERS IN THE FIELD: COAGULANTS AND ALGICIDES

There are some metallic agents that are used in the water purification processes themselves. These are ferric chloride and related compounds, alum and related aluminum compounds, and copper sulfate. All of these agents are somewhat acidic. In addition to the primary metal toxicity, they can burn the skin quite severely. If the agents are shipped dry, there can be a dust problem created by their handling. For all of these agents, the operator should be advised to wear protective goggles, gloves, and mask. Separate work clothes should be kept just for handling these agents. The employer should also provide shower, laundry, and changing facilities for the operators. Thus, the worker can shower and completely change after handling the chemicals. This greatly reduces the chance of inadvertent skin damage due to accidental exposure.

Copper Sulfate — $CuSO_4$

This metal compound is a common algicide. It is toxic to benthos (bottom-dwelling organisms) and other aquatic life, as well as to us. Copper sulfate should be used sparingly for algae control. It would be better to control algae blooms by controlling the cause, that is, limit the amount of phosphorus going into a water body. If this agent must be used, the dosage should be no greater than 1.0 mg/l.[10] The toxicity of copper sulfate is dependent on pH and alkalinity. So bioassay studies may be necessary to determine the safe dosage for a particular water body. There is also information available to help the operator determine safe dosages in his or her area.

Even though copper is toxic, it is needed in small amounts as a nutrient. The daily dosage from whole grains, leafy vegetables, legumes, and liver is from 2.5 to 5.0 mg.[7] Since we get all the Cu we need from our diet, it is certainly not necessary to get it at work. Copper sulfate is called blue vitriol because of its ability to burn tissue. TLVs for Cu are 1.0 mg/m^3 as a mist and 0.2 mg/m^3 as a fume.[3] Handling copper sulfate powder can create respiratory problems. Also, copper can cause ulceration and perforation of the nasal septum. Digestive disorders can result over a period of time, and the worker's skin may actually take on a greenish hue.[3] Use a mask, gloves, and goggles when using the chemical.

If the salt gets in the eyes, flush the area with water for 20 min. Even if it appears the eye is not damaged, take the person to a qualified ophthalmologist. All eye injuries require the very best care. Skin burns should be flushed with water for 15 min. Do not attempt to neutralize the area. Clothing exposed to copper compounds should be thoroughly laundered. Badly saturated clothing, of course, is unsafe and unusable. Store copper compounds in a cool, dry place in acid- and corrosion-resistant containers. Copper sulfate is not as toxic as some previously discussed metals, but it does require respect in its handling.

Ferric Chloride and Related Salts — $FeCl_3$

The ferric salts are widely used as agents for phosphorus and solids removal. Besides ferric chloride, ferrous sulfate (copperas) and a mixture of ferrous sulfate and ferrous chloride, called chlorinated copperas, are used in conjunction with lime. Of all the chemicals thus far discussed, the iron compounds seem to have the least toxicity. Iron is also a necessary nutrient. Requirements for this substance range from 15 to 30 mg/d. A well-rounded diet including eggs, red meats, and leafy vegetables will supply man with all the needed iron.

Although the primary toxicity of iron is small, it still requires careful handling, as most of the salts and their solutions are quite acidic. The author has been burned by handling ferric chloride and can attest to the compound's ability to attack cloth and stainless steel. Iron salts can be ordered in 55-gal drums or even in bulk shipments for large treatment plants. Liquid ferric chloride is commonly shipped as a 30 to 40%

solution, because the salts are so deliquescent. They readily adsorb water from the air. If ferrous sulfate is shipped in its dry state, it must be stored in a dry and cool place to avoid caking. All of these salts will attack stainless steel and other metals. The containers and pumps used in handling these salts must also be chemically resistant.

If ferric coagulants are splashed in the eye, one must flush the eye 20 min with clean water. Skin burns are simply flushed for 15 min with water, and the injury is given first aid for burns. Call 911 after first aid is applied. Avoid stirring up a mist of these salts, as their TLV is 1.0 mg/m^3 as Fe. Spills should be cleaned up promptly to avoid damaging equipment. Hose the area with copious amounts of water.

Alum and its Related Salts — $Al_2(SO_4)_3 \cdot 14H_2O$

Alum and its related salts[8] are also used extensively as coagulants. The choice between alum and ferric salts may be due to the properties of the plant's raw water. The pH, alkalinity, and other conditions must be determined, and jar tests often indicate which coagulant is best. Some operators claim alum is easier to handle than ferric chloride. Even though this agent is acidic, it does not seem to have the same degree of corrosivity as the iron salts. It is recommended that the dust from handling the dry powders be kept below a TLV of 2.0 mg/m^3. In treating alum burns, the affected area should be flushed for 15 min with water. One can then apply a paste of sodium bicarbonate for 1 min, and then flush the skin for another 5 min. If the eyes are affected, they should be flushed for 20 min with plain water. An ophthalmologist should make sure no permanent damage has occurred.

Both the dry aluminum salts and their solutions should be stored in chemically resistant containers in a dry, cool area. Alum and its related compounds are incompatible with sodium chlorite. Their combination can result in an explosive formation of sulfur oxides and chlorine. Care should be taken to see that the alum storage facilities are kept completely separate from disinfection storage facilities.

Acute injury by alum salts is treated as any acid burn, as stated before. However, the chronic toxicity of aluminum salts is not well understood and is still debated by the medical community. Over the last 10 years some disturbing epidemiological research has been done concerning long-term exposure of workers and other people to chronic doses of aluminum and related salts. There may be evidence that aluminum exposure may induce Parkinson's disease.[11] Certain people may not be able to excrete excess aluminum out of the body as effectively as others. The life burden of aluminum in most normal individuals may be 50 to 150 mg per man.[7] Although it has no known nutritive value, aluminum is found in many foods and is listed as an ingredient in many vitamin and mineral formulas. The degree to which aluminum is adsorbed seems dependent on the form of aluminum, the age of the exposed person, the quality of diet, and the possible genetic disposition. For instance, aluminum hydroxide has a solubility of 1×10^{-32} K_{sp} at 25°C in a neutral pH.[12] But $AlCl_3$ or $Al_2(SO_4)_3$ are quite soluble under the same conditions. If conditions are noticeably more acid or basic, even the hydroxide can become soluble. Also, older people may have a harder time clearing their bodies of excess aluminum. A poor diet can exacerbate the pathology of aluminum poisoning. A good diet should be rich in B complex; vitamins A, C, and E; calcium; and magnesium. Calcium and magnesium seem to be especially important in giving the worker some protection from alum and its related salts.[13] In addition to aluminum being implicated in Parkinson's disease, it seems also suspect in Alzheimer's disease and amyotrophic lateral sclerosis.[13] The diseases mentioned also seem to have strong familial patterns, but there is some evidence that the environment can either cancel or enhance these genetic patterns: witness the high incidence of these diseases in the aluminum rich and calcium- and magnesium-poor areas of Japan, Guam, and New Guinea.[12] Other areas in South America seem to have very low occurrences of these diseases. One chilling positive correlation with aluminum and neural pathology is the incidence of a disease called dialysis encephalopathy found in people undergoing dialysis.[12] This is a gradual degeneration of the nervous system, which will eventually completely disable the person undergoing dialysis. Doctors have now found that the water used in dialysis must be absolutely free of aluminum and other metals, to prevent such side effects. People who work processing aluminum also show other kinds of sickness. These people have liver, kidney, and bone problems, as well as anemia.[7,12]

What does all this have to do with the wastewater and water operator? The point is this: although most references in the water quality field do not mention any special dangers in handling alum salts, it would be wise for the operator to treat this salt with respect. The inhalation of alum dust could possibly be a direct route for aluminum's entrance into the bloodstream. Even if there is not a direct correlation of aluminum with the aforementioned conditions, the operator would be wise to wear a protective respirator when handling these salts in a dry condition. After a day of handling coagulant salts, the operator should shower and completely change from the clothing worn at work. If aluminum poisoning is suspected at all, the worker should have his blood and hair analyzed for this metal. Self treatment with chelation agents should NEVER be attempted, since most chelation agents have dangerous side effects. This is the province of an environmental specialist in medicine.

Toxicity of Metals in General

Only a few of the major metals have been covered. It is good for the environmentalist to remember that all metals can be toxic in excessive doses. Even nutrient metals can cause problems if they get into the respiratory system. It is also important to remember that many metals can have similar pathologies: witness mercury, lead, manganese, and aluminum. All of these metals can induce neurological symptoms of some sort. In addition to some metals having similar pathologies, the symptoms resulting from exposure can mimic other unrelated diseases. It is therefore important that all suspect agents be handled with care. It is also important that the employer be vigilant for any unusual symptoms or behavior in his workers. Is that ill-tempered worker just a sullen personality or is that lead poisoning talking? The key in good industrial hygiene is to be aware of CHANGES in behavior. Once a problem is suspected, a worker should be encouraged to see an environmental specialist. It distresses the author to see that many general practitioners are ignorant of the workplace factors that might affect the health of the patient. Health professionals should always be aware of the vocational status of their patients. Medical histories should include vocation and a list of potential hazards. All doctors need to keep up with the latest information in industrial safety. Environmental workers are examples of the type of patients who have special health needs due to the potential hazards of their occupation.

Table 3.1 offers a short summary of those metals most important to the water quality field. This is by no means a complete list. Any chemical should be thoroughly researched before being used in the plant.

Table 3.1 Metals and their Properties

Metal	Source	Pathology	Precautions and First Aid
Aluminum (Al)	Coagulant salts	Nervous system, kidneys, liver, bone	Never store with sodium chlorite; protect lungs, eyes, and skin; flush exposed area with water
Beryllium (Be)	Pollution samples	Lung granuloma, severe contact dermatitis	Positively no contact with this metal without glove boxes and ventilation; flush exposed area with water

Table 3.1 Metals and their Properties (continued)

Metal	Source	Pathology	Precautions and First Aid
Cadmium (Ca)	Pollution samples	Bone disease, kidney damage	Use gloves, eye protection, and ventilation; thoroughly flush affected area
Chromium (Cr)	Oxidants, cleaning agents, pollution samples	Cancer, contact dermatitis	Use gloves, eye protection, and ventilation; thoroughly flush affected area; neutralize with $NaHCO_3$ and flush again; treat as severe burn; don't store salts with organics
Copper (Cu)	Algicide	Burns, lung damage	Gloves and eye protection; flush affected area
Iron (Fe)	Coagulant salts	Burns, lung damage	Gloves and eye protection; flush affected area
Mercury (Hg)	Reagents, pollution samples	Mainly central nervous system damage	Ventilation of utmost importance; protect skin; flush affected area; be alert for changes in worker behavior
Manganese (Mn)	Taste and odor control, Fe and Mn removal	Central nervous system, lungs, and skin; obsessive and compulsive behavior	Protect lungs, eyes, and skin; flush affected area with large amounts of water; ascorbic acid will remove stains on skin, but rinse thoroughly; don't store with organics
Lead (Pb)	Reagents, pollution samples	Central nervous system, bones, blood	Protect lungs, eyes, and skin; flush affected area

REFERENCES

1. Steere, N.V., Ed. 1971. *CRC Handbook of Laboratory Safety*. 2nd ed. CRC Press, Boca Raton, FL.
2. Carroll, L. 1863. Alice's Adventures in Wonderland.In*The Complete Works of Lewis Carroll*, 1st ed. Random House, New York. p. 79.
3. Proctor, N.H., and J.P. Hughes. 1978. *Chemical Hazards of the Workplace*. J.B. Lippencott, Philadelphia.
4. Waldbott, G.L. 1978. *Health Effect of Environmental Pollutants*. C.V. Mosby, St. Louis.
5. Nriagu, J.O. 1983. Saturine gout among Roman aristocrats: did lead poisoning contribute to the fall of the empire? *N. Engl. J. Med.* 308(11):660.
6. Jueneman, F.B. 1983. A lead-pipe cinch, *Ind. Res. Dev.* p. 19.
7. Kirschmann, J.D. 1975. *Nutrition Almanac*. McGraw-Hill, New York. p. 66.
8. American Water Works Association, Inc. 1971. *Water Quality and Treatment: a Handbook of Public Water Supplies*. McGraw-Hill, New York.
9. Standeven, A.M., and K.E. Watterhahn. 1989. Chromium (VI) toxicity: uptake, reduction, and DNA damage. *J. Am. Coll. Toxicol.* 8(7):1275.
10. Machenthun, K.M. 1969. *The Practice of Water Pollution Biology*. Federal Water Pollution Control Administration. p. 240.
11. Zayed, J. et al. 1990. Environmental contamination by metals and Parkinson's disease. *Water Air Soil Pollut.* 49:117.
12. Sawyer, C.N., and P.L. McCarty. 1978. *Chemistry for Environmental Engineering*. McGraw-Hill, New York.
13. Weiner, M.A. 1987. *Reducing the Risk of Alzheimer's*. Stein and Day, Briarcliff Manor, NY.

4
Chemical Hazards — Organics

4.1 SOURCES OF CONTACT

There are several organics commonly used in the course of a day in water and wastewater treatment plants (Table 4.1). Ketones, ethers, and freons are used for extraction procedures. Some alcohols, halogenated hydrocarbons, alkanes, and nitriles are used in gas chromatography (GC), liquid chromatography (LC), and infrared work (IR or FTIR) as preparation and cleanup agents. The standards used in such work are usually a combination of organophosphates, carbamates, and ring-structure herbicides and fungicides. These and other priority pollutants are highly toxic and require the utmost caution in their preparation and handling.

The lab is not the only area where a worker can be exposed to organic agents. Solvents are often used in painting and for cleaning heavy equipment. Chlorinated organics, and organophosphates, are sometimes used to control filter flies and other pests around the plant. Certain herbicides are used in keeping the grounds neat. So the operators or laborers are often in danger of contamination from these agents.

People are often careful to protect their lungs from such agents, but do not always realize that the skin is a major route of exposure. Nor do workers realize just how toxic these agents can be. For instance, organophosphates have detrimental effects on the central nervous system (CNS) at doses of hundredths to tenths of milligrams per kilogram of tissue. A worker may not realize he has been splashed with a toxic agent until symptoms appear. TLVs are often set at less than 0.1 mg/m^3. Therefore, it is important for the worker to be properly clothed for work with such agents. Training is a must. Many organics require not only protective clothing, but also special handling.

42 CHEMICAL HAZARDS AT WATER TREATMENT PLANTS

Table 4.1 Dangerous Organic Compounds

Terpenes	Halogenated hydrocarbons	Halogenated cyclics and aromatics
(structure with CH_3, CH_3, CH_3)	$H-\underset{H}{\overset{R}{C}}-X$ $H-\underset{H}{\overset{X}{C}}-X$ $X-\underset{X}{\overset{X}{C}}-\underset{X}{\overset{X}{C}}-X$ X = F, Cl, Br, or I	(phenyl)−X (cyclic)−X X = F, Cl, Br, or I

Thiols	Phenols	Phosphates and thiophosphate esters (thions and oxones)
R−SH $R-\underset{H}{\overset{R'}{\underset{S}{C}}}-R'$ $R-\underset{H}{\overset{R'}{\underset{S}{C}}}-R''$	(phenyl)−OH (R-phenyl)−OH	$(RO)_2-\overset{S}{\underset{\|\|}{P}}-OR'$ $(RO)_2-\overset{O}{\underset{\|\|}{P}}-OR'$

Alcohols	Carbamates	Pyrethrums
R−OH $R-\underset{OH}{\overset{R'}{C}}-R'$ $R-\underset{OH}{\overset{R'}{C}}-R''$	$RO-\overset{O}{\underset{\|\|}{C}}-N\underset{R}{\overset{H}{\diagdown}}$	(structure) R or X, R or X X = F, Cl, Br, or I

Bipyridyl compounds
$R-\overset{+}{N}$(phenyl)−(phenyl)$\overset{+}{N}-R'$

4.2 HEALTH EFFECTS OF ORGANICS

There are a bewildering array of organic chemicals. Rather than attempt to cover organic chemicals on the basis of structure, it would be more useful for the purposes of this work to group the chemicals based on their effect on the body. Where appropriate, specific organics will be mentioned in the context of their usefulness or threat to the water quality field.

There are three major ways organics can affect the body. The most common threat is to the CNS. There are two major submodes of destruction to the CNS. The first and most common is the direct destruction of nerve tissue. Ketones, ethers, aliphatics, and aromatics can cause pathology in coordination and emotions by attacking the cerebellum and the frontal lobes of the brain. The other mode whereby the CNS can be disrupted results from the agent attacking specific enzymes needed for the smooth functioning of sensory organs. Organophosphates are notorious for "binding" the enzyme acetylcholinesterase. This allows

CHEMICAL HAZARDS — ORGANICS

Malathion

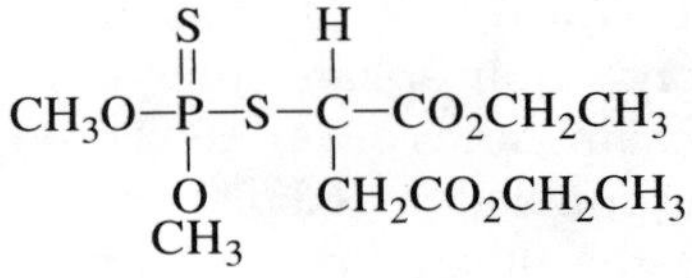

$$CH_3O-\underset{\underset{CH_3}{O}}{\overset{\overset{S}{\|}}{P}}-S-\underset{CH_2CO_2CH_2CH_3}{\overset{H}{C}}-CO_2CH_2CH_3$$

Parathion

$$CH_3CH_2O-\underset{OCH_2CH_3}{\overset{\overset{S}{\|}}{P}}-O-\!\!\left\langle\bigcirc\right\rangle\!\!-NO_2$$

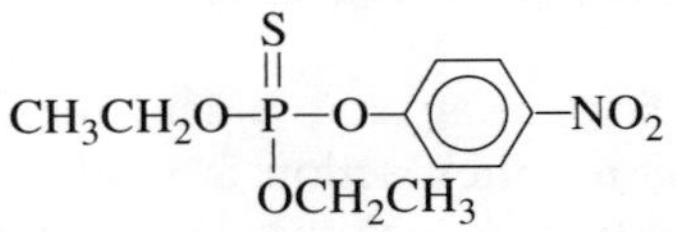

In the course of toxic action, Thion is oxidized to oxone

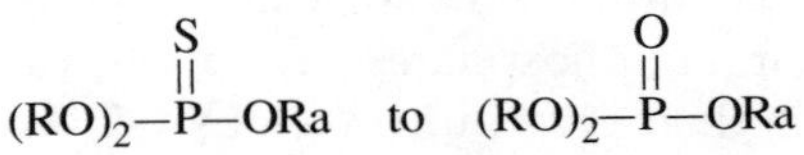

$$(RO)_2-\overset{\overset{S}{\|}}{P}-ORa \quad \text{to} \quad (RO)_2-\overset{\overset{O}{\|}}{P}-ORa$$

Figure 4.1 Organophosphate structure.

Some workers become dependant on the fumes of solvents such as ethyl ether or methyl-ethyl ketone. Last, but not least, are the carcinogens. Many different kinds of aliphatic and aromatic hydrocarbons can induce cancer after a long period of chronic exposure. There are also reproductive effects from organics, but since birth defects and their causes are such a broad topic, this aspect of organic exposure will be covered in a separate chapter.

The Nerve Agents: Organophosphates

Organophosphates are used both in the field and in the lab. It is important to be skilled in their handling, because careless usage can result in violent and fatal convulsions in a short time. In the lab they are commonly used in a purified form as standards for gas chromatography. Because they are in such a pure form, the first dilutions of standards are usually done under a self-contained glove gas hood.

The groundskeeper will often use a common malathion- or parathion-based insecticide for control of such pests as filter flies or mosquitoes. Quite often a worker using an insecticide will not give any thought to protective clothing. It is the author's opinion that even home-use products are not nearly adequately labeled as to their handling and first aid requirements. It is puzzling to see an operator, normally so careful about a coagulant or an oxidant, handle a pesticide is such a casual manner. Folks! The granddaddies of these agents WERE the nerve agents used in World War II.[1] That fact should never be forgotten.

The structure of organophosphates may either have aromatic ring structures, aliphatic chains, and/or carboxyl groups. One thing they all have in common is the double-bonded phosphate-sulfur structure called a thion group. Note the illustrations (Figure 4.1) of some common pesticides. It is this group that is oxidized to the oxone group, and this is the structure that binds so easily to the cholinesterase enzyme.[2] This bonding is more or less irreversible, so it only takes a little of the toxins to injure a worker.

Because the poisons attack the biochemical pathways involved in nervous transmission, quick action is vital if a worker's life is to be saved. The exposed worker's clothes should be removed immediately, and the skin washed thoroughly with soap and water. Flood the worker's skin for at least 20 min. Use any available source of water to get the substance off the skin quickly. Destroy all exposed clothing, including belts and shoes. Organophosphates cannot be washed out of leather. Atropine sulfate is then given intravenously ONLY if a fully trained company doctor is on the premises. If the treatment plant has no resident nurse or doctor, call 911 and have an ambulance rush the person to the nearest hospital. Make sure the medics know exactly what insecticide was involved in the injury. Call ahead to the hospital with the exact composition of the substance. If an MSDS sheet is availible, have that also taken to the hospital. Time is of the essence, and the information provided can spell the difference between complete recovery, permanent disability, or death. The hospital will usually keep the injured worker for 24 to 48 h after the incident to be certain that all traces of the poison have been neutralized. Once a person has been injured with an organophosphate, he remains more susceptible to the poison for quite some time. The body is only able to synthesize new enzymes at a rate of 1% per day.[3] Thus, the recovered operator should take special care to avoid future exposures.

The salivation, lacrimation or tearing, urination, dizziness, diarrhea, pallor, nausea, and convulsions of an acute episode are easily recognized symptoms; most people only think of exposure to the organophosphates in terms of acute pathology. Yet there is evidence that exposure to minute amounts of these substances, over time, can cause chronic problems later. Also, persons recovered from an acute incident will have some lingering nervous impairment. As little as 0.05 ppm of organophosphate residue in urine can be associated with complaints of dizziness and impaired vision.[4] But since individual tolerances vary greatly between individuals, it is difficult to give any clear picture of a typical chronic toxicity from these pesticides. New evidence is showing that subtle changes in the intellectual ability and emotional stability of workers could be the result of previous acute episodes or chronic exposure.[5] Symptoms range from inability to use language correctly or to think abstractly, to depression and irritability. Gross motor damage is not a problem. But since most doctors only look for coordinational damage, the emotional and intellectual damage caused by organophosphate exposure is often overlooked.[5] What this means is that the employer should carefully observe a previously injured worker. Any changes in work performance or emotional stability that occur after exposure to organophosphates should be taken seriously.

It is quite apparent that the organophosphates are deadly pollutants. Strict rules should govern their handling. These general guidelines should be helpful:

1. In the lab, all personnel should wear gloves when working with GC standards.
2. No standard should be prepared unless an enclosed gas hood is provided.
3. After each lab session the work area and all glassware should be completely scrubbed down.
4. The lab worker should thoroughly scrub hands for 5 to 10 min after work on the GC.
5. In the field, choose only clear, windless days for applying a pesticide to a problem area.
6. Wear a respirator, goggles, old clothes, and long rubber gloves when mixing a preparation.
7. Only mix a preparation in a WELL-VENTILATED area.
8. Follow the manufacturer's directions exactly.
9. Make sure any spillage is completely scrubbed with a strong detergent and hosed down.
10. After the application is complete, remove your old clothing and take a vigorous shower. Make sure all parts of the body are scrubbed and rinsed before drying off and changing to clean clothing. (Most modern treatment plants provide shower facilities for their personnel.)
11. Groundskeepers should be tested periodically by their doctors for red blood cell cholinesterase levels. Workers with abnormally low levels should be protected from further exposure to these agents.
12. DO NOT store these insecticides with any other oxidants. The byproducts of decomposed pesticides can be worse than the original product.[6]

Nerve Agents: Methyl Carbamates

Methyl carbamates such as Sevin® dust may be used in place of the organophosphates. Note the structure of a basic methyl carbamate in

Sevin or Carbaryl

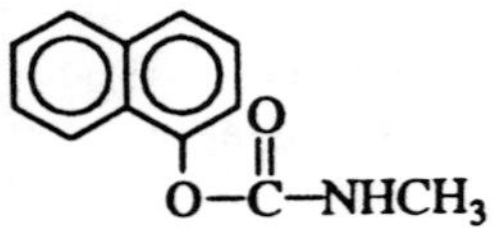

Aldicarb

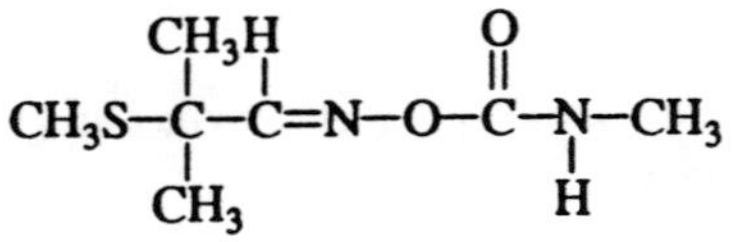

Figure 4.2 Methyl carbamates.

Pyrethrin I

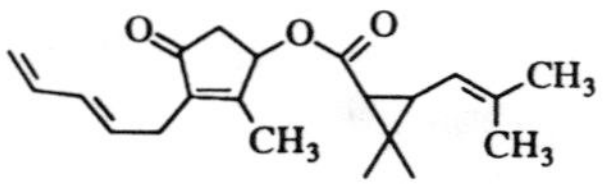

Permethrin

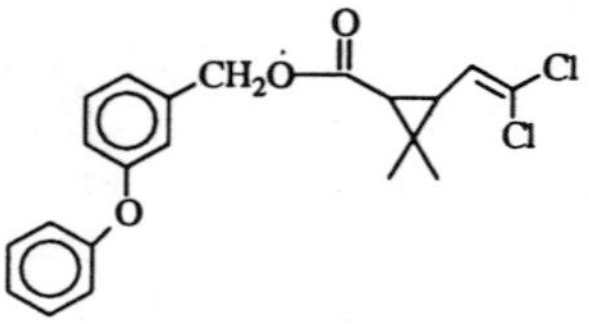

Figure 4.3 Pyrethrum products.

Figure 4.2. They also bind the acetylcholinesterase enzymes, but not as efficiently as do the organophosphates. They are probably synergistic with the organophosphates, that is, they can worsen nervous damage if used with these agents. DO NOT use organophosphates and carbamates together. The safety and first aid directives for carbamate insecticides are the same as those for organophosphates. The key in using both of these agents is cleanliness. A worker must insist on proper handling and good hygiene.

CHEMICAL HAZARDS — ORGANICS

Paraquat

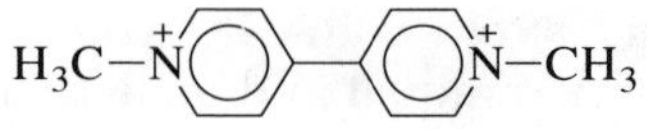

$$H_3C-\overset{+}{N}\bigcirc-\bigcirc\overset{+}{N}-CH_3$$

Diquat

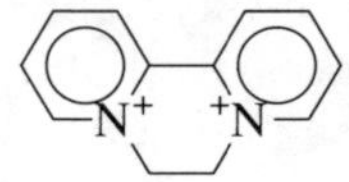

Figure 4.4 Bipyridyl compounds.

A Possibly Safer Substitute

For most grounds problems, one should consider the use of pyrethrum-based insecticides (Figure 4.3). This substance has none of the severe neural effects of the organophosphates or the carbamates. The pyrethrums are a natural, concentrated extract of repellent chemicals found in marigold flowers. There are still a few precautions one must take with this substance, however. Some people are severely allergic to the marigold family of flowers. This means they are also allergic to pyrethrums. Also, the pyrethrum poisons are especially potent to aquatic life. These insecticides would be a poor choice to control filter flies.

One should be careful not to breathe the dust or mist of these products. The TLV of these derivatives is 5 mg/m^3.[2] Contact of the pyrethrums with the skin can cause severe dermatitis. Make sure the operator cleans up any spillage thoroughly. One should also wash and change into clean clothing after working with this substance. Any person with hay fever should not be allowed to work with pyrethrums. If the substance gets on the skin, wash with generous amounts of soap and water. Although this is a mild poison as far as man is concerned, it still deserves to be handled with respect.

Other Metabolism Disrupters: The Bipyridyl Compounds

These are biaromatic compounds with an active nitrate group (Figure 4.4). They are not commonly used by groundskeepers as herbicides, and the author would recommend that their use be avoided. They are potent antimetabolites. They replace niacin in the body. However, the body attempts to use a bipyridayl as if it were niacin. Thus, the growth of the "fooled" cell is disrupted. Weed by hand. These deadly poisons are not worth the effort. Unfortunately, a chemist running the gas chromatograph cannot avoid using the compounds as standards in priority pollutant work. Paraquat and its derivatives only have a TLV of 0.5 mg/m^3.[3] This value is with good reason. These herbicides cause irreversible damage to the lungs, liver, and kidneys. Dosages less than 40 mg/kg can be immediately fatal.[2] There is no antidote except for good supportive hospital care. Since oxygen therapy might enhance this poison's action, only medically trained personnel should deal with an injury from these substances. Even if the worker survives, fibroblastic tumors in the lungs cause permanent disability.[3,7]

These compounds require an enclosed gas hood for standard preparation. The chemist should be very careful to protect the skin with gloves. After a determination, the lab area should be meticulously scrubbed clean. If the chemical is accidentally splashed on the skin or in the eyes, flush the areas with water for 20 min. Call 911 for an ambulance to take the person to the nearest hospital. Even with good supportive care, some permanent damage to the lungs may be inevitable. Chronic congestion, fibrosis, and thickened arteries are the result of chronic low level exposure to bipyridyl compounds.[7] An acute episode can lead to respiratory failure and death within 2 weeks of exposure.[7] Remember, this class of compounds is skin active. They pass the skin barrier easily and require great caution in their preparation as analytical standards. Never forget to wear gloves.

It would be beyond the scope of this book to describe the toxicity of all of the priority pollutants. It serves to remember that most of these organic compounds are easily absorbed through the skin as well as the lungs. Most have addictive, neurological, or systemic effects on the body. Most are quite toxic in very small amounts. This should be borne in mind as the chemist prepares highly purified extracts for analytical work. Observing safe laboratory policy is the first and most effective line of protection.

Addictive Agents and Nervous Deterioration

Many kinds of organics can cause addictive behavior and slow deterioration of the CNS. Alcohols, ketones, halogenated hydrocarbons, and many other kinds of hydrocarbons can become a problem for the worker, both in the field and in the lab. The operators often work with such agents as turpentine, gasoline, or xylene in their daily routine. Chemists work with a bewildering variety of solvents and carriers. The single most important protection that could be offered to any worker is that of good ventilation and skin protection. The effects of most of these agents manifest themselves very slowly over a period of time. Avoid their misuse from the start.

CHEMICAL HAZARDS — ORGANICS

In maintaining equipment and grounds, the operator or laborer is often exposed to gasoline and solvents. Common paints and varnishes may have toluene and xylene bases. Gasoline is a mix of hydrocarbon chains and aromatics. Turpentine is a mix of a class of organics called terpenes (Figure 4.5).[3] All of these can cause central nervous system pathology. Most are highly flammable and require that no open flames or sparks be used in an area where they are handled. Nearly all organics, especially motor oils and gasoline, can react violently with chlorine and its related disinfectants. The end product of such an unfortunate mix is a mustard-gas-like substance. DO NOT store any solvent or petroleum product with strong oxidants! This mistake is made at work and also at home. For example, the average home pool owner thinks nothing of storing motor oil alongside the calcium hypochlorite used in the family pool. Remember that organics require ventilation when used. They also require their own cool, dry storage area away from direct sunlight.

Since the aromatics, such as xylene and toluene, are most likely encountered in the workshop, they will be discussed as field agents (Figure 4.6). Both toluene and xylene have TLVs of 100 ppm.[3] This is cause for some concern because chronic poisoning can progress gradually over a long period of time. Yet no overt CNS effects have been reported for toluene until 200 ppm.[8] Both compounds can cause kidney and liver damage. Toluene has also been documented as an addictive substance.[3,8] Both toluene and xylene can cause gradual deterioration of the CNS. Workers who have worked with these solvents for years may have a slapping, shuffling gait as a result of CNS damage. Both agents have also been suspected of causing blood disorders. Mild anemia has been seen in people who have used paint solvents for years.[8] Acute episodes of poisoning can occur at levels of 600 to 10,000 ppm. Dizziness, incoordination, digestive upsets, lung and eye irritation, and loss of consciousness can result from these levels.

Turpentine is primarily an irritant. It can actually burn the skin. Some people experience intense irritation of the eye if they are exposed to the fumes for very long. The TLV for turpentine is 100 ppm.[3] Other effects are CNS deterioration and kidney damage. Some people, when exposed to turpentine, will have allergic reactions. Turpentine can cross-sensitize a person to other organics with similar terpene structures, that is, a person allergic to turpentine will have a reaction to other terpenes when exposed to them for the first time.

With any of these common solvents, one should be careful only to use them in well-ventilated areas. Wear resistant gloves. Use eye protection. Make sure all spills are scrubbed clean with strong detergent. If a solvent is splashed on the skin or eyes, flush the affected areas for at least 20 min with clean water. The skin may be scrubbed with a strong soap and rinsed thoroughly. No other neutralization should be

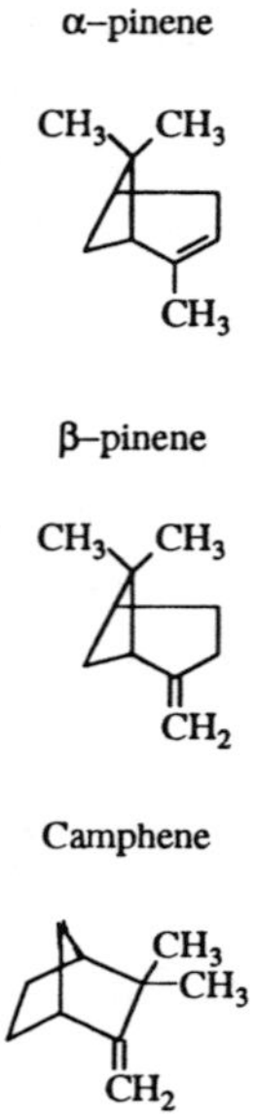

50 CHEMICAL HAZARDS AT WATER TREATMENT PLANTS

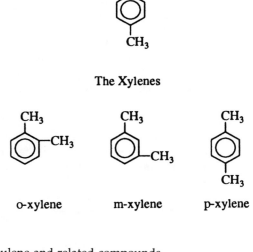

Figure 4.5 Terpenes ($C_{10}H_{16}$).

Figure 4.6 Toulene and related compounds.

attempted. If a person is overcome with fumes, remove him from the area. Initiate artificial respiration or cardiopulmonary respiration (CPR) if needed. All persons overcome with fumes should be put under hospital observation for at least 24 h, regardless of how they say they feel. People with a history of allergies should be especially careful in the handling of solvents.

Addiction is possible with the solvents, especially toluene. Most "glue heads" show reduced alertness, increased nervousness, insomnia, and a lack of coordination. Addicts will also show metabolic acidosis because of the body's attempt to break down the poison. Liver, kidney, and blood damage can also occur in addicts.[8] As with alcoholism and other addictive behaviors, it is not always easy for the superintendent to spot workers who may be in need of help. The key, again, is to be on the alert for changes in behavior over a period of time. Disciplinary and rehabilitative action should be taken. Any person with a history of addiction to any substance should not be allowed to work with chemicals of suspected addictive properties.

Dangers in the Lab

The water quality chemist is not immune to addiction and nervous deterioration. In fact, in many ways their situation may be worse because the chemist may work continually with these agents, as a specialist. Plant personnel may work with solvents only occasionally.

Some of the agents used to separate or extract samples for analysis are the worst offenders. Ethyl ether and methyl isobutyl ketone are commonly used as cleaning agents or separating agents in GC or AA work. Both are powerfully addictive, highly flammable, and neurologically toxic. Because they are used frequently in the pollution field, they will be covered separately.

Methyl isobutyl ketone has the structure seen in Figure 4.7. This substance and other ketones are used to separate heavy metals from extremely oily samples before atomic adsorption analysis (AA). Ketones are addictive for certain individuals. Methyl isobutyl ketone has a TLV of 100 ppm.[3] At 25 ppm most people can detect a ketone odor from these chemicals. Exposure over a number of years can damage the CNS. Workers exposed to around 400 ppm complain of irritation to the eyes and nose.[3] There is some evidence that exposure to some ketones could cause behavior problems such as difficulty in concentration and memory.[9] The TLV of 100 ppm is questionable, as some individuals can be affected by 100 ppm of methyl isobutyl ketone.[3] Another common ketone used in the lab is acetone. It has a TLV of 1000 ppm.[3] Although

Acetone

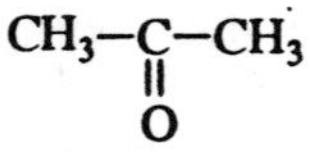

$$CH_3-\underset{\underset{O}{\|}}{C}-CH_3$$

Methyl Ethyl Ketone

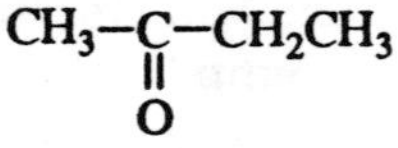

$$CH_3-\underset{\underset{O}{\|}}{C}-CH_2CH_3$$

Methyl Isobutyl Ketone

$$CH_3-\underset{\underset{O}{\|}}{C}-\underset{\underset{H}{|}}{\overset{\overset{H}{|}}{C}}-\underset{\underset{H}{|}}{\overset{\overset{CH_3}{|}}{C}}-CH_3$$

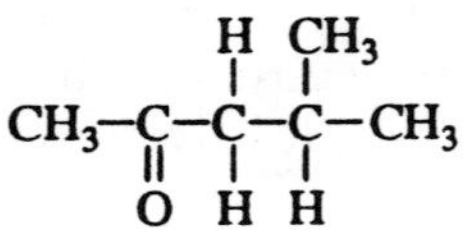

Figure 4.7 Ketones.

a TLV of 100 ppm.[3] At 25 ppm most people can detect a ketone odor from these chemicals. Exposure over a number of years can damage the CNS. Workers exposed to around 400 ppm complain of irritation to the eyes and nose.[3] There is some evidence that exposure to some ketones could cause behavior problems such as difficulty in concentration and memory.[9] The TLV of 100 ppm is questionable, as some individuals can be affected by 100 ppm of methyl isobutyl ketone.[3] Another common ketone used in the lab is acetone. It has a TLV of 1000 ppm.[3] Although milder in toxicity, it can still have some addictive qualities and requires respect in handling.

The key to safe handling of ketones is, again, good ventilation. Never work without a gas hood. Avoid splashing ketones on the skin or eyes. They can dry and irritate skin severely, and damage mucus membranes. Here again, some people also show allergic skin reactions to ketones. Any exposed areas should be washed thoroughly with water. Never attempt to neutralize an organic with another chemical, as this could enhance adsorption through the skin. Just get it off the skin as quickly as possible. Employers also should be on the lookout for addictive behavior at work.

Ethers are used in gas chromatography work as sample cleanup reagents. Ethyl ether is the most common of these. See the structure in Figure 4.8. Its TLV is 400 ppm.[3] Here again, some sensitive individuals

Ethyl Ether

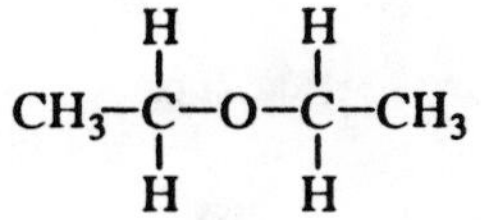

$$CH_3-\underset{\underset{H}{|}}{\overset{\overset{H}{|}}{C}}-O-\underset{\underset{H}{|}}{\overset{\overset{H}{|}}{C}}-CH_3$$

Methly tert-butyl Ether

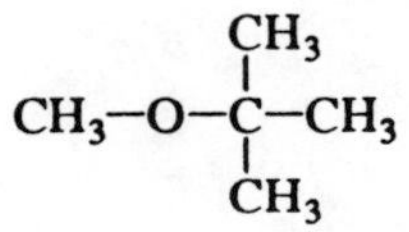

$$CH_3-O-\underset{\underset{CH_3}{|}}{\overset{\overset{CH_3}{|}}{C}}-CH_3$$

Figure 4.8 Ethers.

can have trouble with ether at 200 ppm. This agent is an anesthetic, and workers can easily be overcome. As with any acute episode, the injured party should be immediately removed from the area. Resuscitation should be initiated immediately if needed. This substance is addictive. Chronic exposure leaves a person without an appetite, depressed, and with coordination problems. Ether spills can dry the skin severely. After thorough washing of the skin for 5 to 10 min, one may want to use a moisturizing lotion on the hands. Make sure the hands are completely free of any chemical before using a cream. Lotions and moisturizers may enhance the absorbance of chemicals through the skin barrier and should be used with caution. Ether is also extremely flammable and reactive. It has a tendency to form reactive peroxides upon aging. Ether's explosive properties will be discussed further in a special section dealing with reactive dangers. Ether must be stored in a dark glass bottle and kept in a cool, dry place. Only order enough ether for 4 to 6 months of work. Discard any ether over 6 months old. Always make sure the work area is well ventilated and that there are no open flames or sparks. This agent requires respect, and an ounce of prevention can prevent a pound of troubles.

Hexane and its related compounds are sometimes used as carriers in GC (Figure 4.9). It does not appear to be addictive, but prolonged exposure can cause deterioration of the peripheral nerves. Hence, workers may experience progressive difficulty in maintaining balance and in walking. Its TLV is 100 ppm.[3] It can also irritate the skin, eyes, and nose. Irritating effects occur after 500 to 1000 ppm.[3] This compound is flammable, but it does not form any reactive byproducts upon aging. Be sure

54 CHEMICAL HAZARDS AT WATER TREATMENT PLANTS

n-Hexane

$$CH_3CH_2CH_2CH_2CH_2CH_3$$

Isobutane

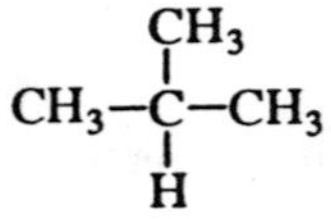

$$CH_3-\underset{\underset{H}{|}}{\overset{\overset{CH_3}{|}}{C}}-CH_3$$

Figure 4.9 The Alkanes.

Methanol

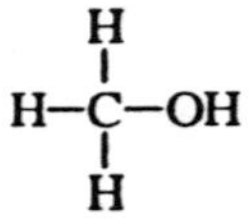

$$H-\underset{\underset{H}{|}}{\overset{\overset{H}{|}}{C}}-OH$$

Ethanol

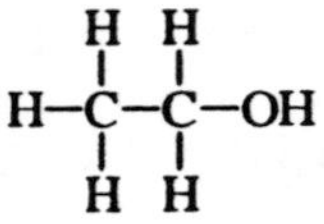

$$H-\underset{\underset{H}{|}}{\overset{\overset{H}{|}}{C}}-\underset{\underset{H}{|}}{\overset{\overset{H}{|}}{C}}-OH$$

Isopropyl Alcohol

$$H-\underset{\underset{H}{|}}{\overset{\overset{H}{|}}{C}}-\underset{\underset{OH}{|}}{\overset{\overset{H}{|}}{C}}-\underset{\underset{H}{|}}{\overset{\overset{H}{|}}{C}}-H$$

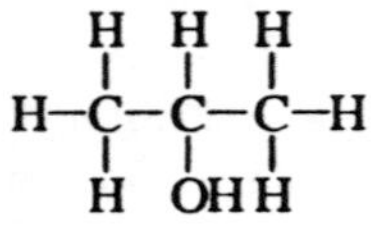

Figure 4.10 Three commonly used alcohols.

that it is stored in a cool, dry area. Glass is the prefered storage vessel.

A variety of alcohols are used as reagents and cleaning agents in the lab. The three most common alcohols are methyl, ethyl, and the propyl alcohols (Figure 4.10). Methyl alcohol has a TLV of 200 ppm, and isopropyl alcohol has a TLV of 400 ppm.[3] Both of these alcohols are skin active. Substantial amounts can enter through the skin. Unlike the mildly toxic ethyl alcohol, methyl and isopropyl alcohols, even in small amounts, can cause severe nervous damage. Methyl alcohol can damage the optic nerves. Isopropyl alcohol is suspected of causing a type of sinus cancer.[3]

When handling these substances, make sure the work area is well ventilated. Protect the skin and eyes from contact. When work is complete, make sure the work area is clean. Wash the hands thoroughly. Any worker showing dizziness, nausea, headache, giddiness, eye problems, or a staggering gait should be rushed to the hospital immediately. The treatment of alcohol poisoning is the province of a trained specialist.

The Systemic Poisons

There are two ways organics can become systemic poisons (that is affecting the whole body). One way is through damage to organs such as the liver, kidneys, skin, and heart. Another way is by causing various types of cancers. Some organics are capable of damaging tissues directly and inducing cancer years later. Some of the agents are also addictive and damaging to the CNS. However, this section will focus on the systemic effects of these particular agents.

There are several agents that can induce cancer after years of chronic exposure. The halogenated hydrocarbons such as chloroform, methyl chloride, 1,1,2-trichloro-1,2,2-trifluoro ethane (freon), are suspected carcinogens (Figure 4.11). The TLVs for these agents are 25 ppm for chloroform, 100 ppm for methyl chloride, and 1000 ppm for freon.[3] In addition to their possible cancer effects, these agents can also cause skin sensitization; liver, kidney, and heart damage; and CNS damage. Chloroform, in particular, can cause gross damage to the liver. Methyl chloride causes kidney and liver damage and has been implicated as a convulsant.[3] Freon is the mildest of the three. Still, it can dry and sensitize the skin. None of these agents can be completely avoided, as they are used as carriers in GC or as separating agents in chemical extractions. Thus, it is imperative to practice good laboratory hygiene when handling such agents.

Another problem, particularly with freon and its related compounds, is that the halogenated hydrocarbons can collect in low-lying areas in a plant or lab, as they are heavier than air. They can crowd out oxygen in these areas. Therefore, forced ventilation is important. Since these compounds are degreasers, they can dry the skin to the point of burning. Be careful to protect the hands from these reagents. Any spills should be flushed from the eyes and skin with water for 15 to 20 min. The skin may be scrubbed with a good soap, but do not attempt to use any other neutralizing agent. Remove and resuscitate overcome workers. Even if the chemist feels fine after the incident, take the person to the hospital.

Chloroform

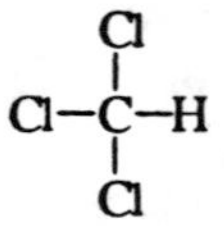

$$\text{Cl}-\underset{\underset{\text{Cl}}{|}}{\overset{\overset{\text{Cl}}{|}}{\text{C}}}-\text{H}$$

Methyl Chloride

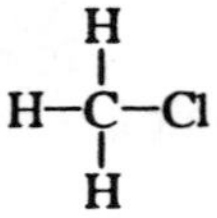

$$\text{H}-\underset{\underset{\text{H}}{|}}{\overset{\overset{\text{H}}{|}}{\text{C}}}-\text{Cl}$$

Freon

$$\text{Cl}-\underset{\underset{\text{Cl}}{|}}{\overset{\overset{\text{F}}{|}}{\text{C}}}-\underset{\underset{\text{F}}{|}}{\overset{\overset{\text{Cl}}{|}}{\text{C}}}-\text{F}$$

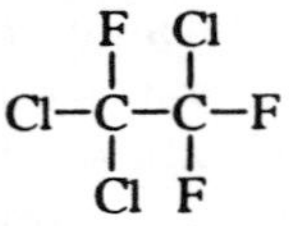

Figure 4.11 Halogenated hydrocarbons.

Phenol

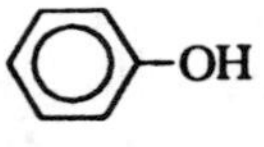

⟨O⟩—OH

Figure 4.12 Phenol.

Methyl chloride, especially, has a latent period of several hours before severe symptoms appear.

Phenol (Figure 4.12) is sometimes used in thin-layer chromatography and certain nitrogen tests. This compound is extremely dangerous and attacks several body systems. Because of its high toxicity and unpredictable behavior, this compound will have a section of its own. Phenol's TLV is 5 ppm, yet some people complain of phenol irritation with as little as 0.05 ppm.[3] This chemical is extremely skin active. It goes directly to the blood from the skin. Because of the peculiar electron structure of the benzene ring, this compound acts more like an acid than an alcohol. In addition to its toxicity, it can cause severe burns. Yet because of its anesthetic effects on the skin nerves, injury may not be discovered until substantial damage has occurred.[10] Once absorbed,

CHEMICAL HAZARDS — ORGANICS

phenol will attack the liver, kidneys, and CNS. As little as 1 g can be rapidly lethal.[3] Chronic symptoms include weight loss, weakness, kidney trouble (as evidenced by dark urine), and an enlarged liver. Very often, workers will develop a sensitivity to phenol and some of its related compounds, after an injury.

Never handle this compound without a gas hood. Be scrupulous in protecting the skin and eyes. Thoroughly clean up the work area after the chemistry work is done. If this agent is splashed on the skin or in the eyes, flush the affected areas with water for 15 to 20 min. Any clothing, shoes, belts, or material that are stained with phenol should be destroyed. It has been documented that clothing can continue to leech poisonous chemicals on the skin.[10] DO NOT attempt to neutralize a phenol burn. DO NOT use any kind of ointment or lotion on a phenol burn. These agents only serve to expand the area of skin exposure. Deaths have been documented from improperly neutralized phenol burns.[10] The best defense against phenol burns is copious applications of clean soap and water. Proctor and Hughes, however, recommend that only water be used to flush a burn.[3] The main thing is that every trace of phenol must be removed from the injury as quickly as possible.

People who have an acute episode of poisoning will show abdominal pain, dark urine, weakness, cyanosis, and tremor. If care is not given immediately, convulsions and death will quickly follow. After cleaning a person up, they should be taken to the nearest hospital. Treatment for phenolic poisoning involves treating the convulsions with sedatives and attempting to stave off liver and kidney damage. Recovery may take weeks. Poisoning requires the care of experts. It is the obligation of the employer to see that any worker injured by phenol follows up with a visit to the hospital or, at the very least, a physician, regardless of how the worker feels. Chronic exposure will result in sensitization of the skin and in chronic liver and kidney disease. Workers who have recovered from chronic exposure sometimes show a hypersensitivity to subsequent exposures to phenol. There is simply no excuse for not providing proper ventilation in the organic chemistry workplace.

Benzene and formalin are implicated in causing several kinds of cancers. They are also systemic poisons in that they can damage the blood, skin, and liver. Benzene is used as a carrier and an extractant in certain chemical tests. Formalin is used to preserve biological specimens for benthic surveys.

Benzene (Figure 4.13) has two safety ceilings: the 1977 TLV is 10 ppm, and the level recommended by OSHA is only 1 ppm.[3,11] The author prefers the 1-ppm level. Benzene has been implicated in causing a type of leukemia and chromosomal damage.[11] Other chronic effects include aplastic anemia, impaired vision, digestive difficulties, and general weakness.[3] Acute injury occurs from 250 to 20,000 ppm. At 250 to 500 ppm, nausea and staggering are seen. The worker may be agi-

58 CHEMICAL HAZARDS AT WATER TREATMENT PLANTS

Benzene

Figure 4.13 Benzene.

Formaldehyde

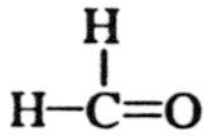

$$\begin{array}{c} H \\ | \\ H-C=O \end{array}$$

Figure 4.14 Formaldehyde.

tated or euphoric. At 3000 ppm, serious damage can occur to the respiratory tract. Death occurs in a few minutes at levels in the 1000s ppm. Benzene is also absorbed through the skin. It can severely irritate the skin. If benzene is splashed in the eyes or on the skin, the areas should be flushed with clean water for 15 to 20 min. Remove and resuscitate overcome individuals. Any person having a benzene injury should have follow-up blood work done by a qualified specialist. Even exposure to small amounts of benzene could do lasting damage to the bone marrow. The safety ceilings have been set to accomodate existing ventilation technology. Remember, there is NO safe level of benzene exposure.[3] Thus, there is positively no excuse for sloppy lab housekeeping. Make sure skin and eyes are protected. Always work with this substance under the gas hood. Diligent attention to detail in lab safety can avert injury and disability.

Formalin, which is a solution of formaldehyde gas (Figure 4.14), has a TLV of 2 ppm.[3] It has been associated with respiratory damage, skin irritation, and cancer. It can burn the skin and cross-sensitize a person to related organic compounds. A severe asthma attack may also result from exposure to formalin fumes.[12] If a person is sensitive to formalin, even a few tenths of a part per million could cause a severe asthmatic reaction. At levels of 50 to 100 ppm, severe pulmonary damage resulting in edema, pneumonia, and death can occur. Splashing a solution of formalin in the eyes can damage the cornea. Any spill should be flushed off the skin or eyes with a steady stream of clean water. Flushing should continue at least 10 to 15 min. Overcome individuals should be removed and resuscitated. As this is a severe pulmonary irritant, the worker should probably be observed at the hospital for at least 24 h.

When working with preserved biological specimens, the benthic specialist should make sure the work area is well ventilated. Wear gloves to protect the skin, as this agent can cause severe dermatitis. Protect the eyes with goggles when making up preservative solutions. After work see that the area is thoroughly cleaned. Wash the hands after working with benthic specimens.

Organics: An Overview

Needless to say, it is impossible to discuss every organic that might be used in the water quality field. It is important to realize, however, that certain general assumptions may be made about all organics for the sake of safety.

1. Assume all organics are skin active; take steps to protect the hands and clothing. Destroy any badly stained clothing. Even leather has been known to leach organic residue days after the initial exposure.
2. Assume all organics can damage the central nervous system, be addictive, or cause systemic disease. Make sure that adequate ventilation is provided in the work area. Clean up all spills. Make sure to leave a clean work area after handling an organic.
3. Assume all organics become more toxic if neutralization is attempted. If a person is contaminated, use large amounts of water to remove the offending substance. Phenol is a notorious example of an organic that can cause more damage when one attempts to use a neutralizing agent without thorough cleaning and rinsing first. Leave neutralization to the poison center of the hospital. Your job is to provide a thorough cleanup and good basic first aid before hospitalization.
4. Assume that all organics can be synergistic if mixed with other agents. PLEASE follow the manufacturer's instructions to the letter when using the organic. This is very important, especially when dealing with pesticides. If there is doubt about how an organic should be used, do not hesitate to call the manufacturer, a local poison center, or another lab experienced in handling the substance.

5. Assume that most organics are violently reactive to acids and oxidants. Keep organics in separate storage areas. Make sure that all reagents carry an expiration date. Properly dispose of any expired organics. This is particularly important with the ethers.

These five assumptions may save a life. Remember, a healthy fear of and respect for organic chemicals can go a long way in averting disaster.

REFERENCES

1. Waldbott, G.L. 1978. *Health Effects of Environmental Pollutants.* C.V. Mosby, St. Louis.
2. University of South Florida. 1984. *Pesticides in Ground Water Symposium.* Tampa, FL.
3. Proctor, N.H., and J.P. Hughes. 1978. *Chemical Hazards of the Workplace.* J.B. Lippencott, Philadelphia.
4. Duncan, R.C., and J. Griffith. 1985. Monitoring study of urinary metabolites and selected symptomatology among Florida citrus workers. *J. Toxicol. Environ. Health.* 16:509–521.
5. Savage, E.P. et al. 1988. Chronic neurological sequelae of acute organophosphate pesticide poisoning. *Arch. Environ. Health.* 43(1):38–45.
6. IOCU. 1986. *The Pesticide Handbook: Profiles for Action.* Organization of Consumer's Union, Penang, Malaysia. p. 153.
7. Revkin, A. 1983. Paraquat a potent weed killer is killing people, *Sci. Dig.* June. pp. 36–42.
8. Haley, T.J. 1987. Toluene: a chemical review. *Dangerous Properties of Industrial Materials Report.* Sept/Oct. pp. 2–14.
9. Anderson, A. 1982. Neurotoxic follies, *Psychol. Today.* July. pp. 30–43.
10. Steere, N.V., Ed. 1971. *CRC Handbook of Laboratory Safety.* 2nd ed. CRC Press, Boca Raton, FL. p. 111.
11. Babich, H. 1985. Reproductive and carcinogenic health risks to hospital personnel from chemical exposure — a literature review, *J. Environ. Health.* 48(2):52–56.
12. Moore, L.L., and E.C. Ogrodnik. 1986. Occupational exposure to formadehyde in mortuaries, *J. Environ. Health.* 49(1):32–35.

5

Caustics and Corrosives

5.1 ACIDS AND BASES

Acid and basic agents are used extensively in both the lab and the field. Common acid agents are hydrochloric, nitric, and sulfuric acids. Common basic or caustic agents are sodium bicarbonate, sodium carbonate, potassium and sodium hydroxide, and calcium oxide and hydroxide. Although the systemic toxicity of these agents is relatively mild, they can cause severe damage to lungs, eyes, and skin by burning. Protection from injury is mainly a manner of proper handling.

All acids have the ability to burn and char the skin. All acid fumes can act as pulmonary irritants. The first aid for injury is relatively straight forward. If acid is splashed on the skin, one should remove clothing from the area and flood the area with clean water for at least 15 min. Then a paste of baking soda, i.e., sodium bicarbonate ($NaHCO_3$), can be applied to the area for about one minute. Then thoroughly flush the soda off with water for an additional 10 min. If the area is blistered, wrap the injury in clean, wet cloth or gauze and take the person to a hospital. The worker should carefully watch the burn area and see a doctor immediately if any complications occur. Eye injury is especially serious. Many labs forbid their workers to wear contact lenses because of the difficulty of removing a lens from an injured eye. Most chemists and operators who work with highly corrosive or caustic agents wear glasses, not contacts. Even with glasses, a face shield or over-goggles should be worn for additional protection. If the eye should be injured, immediately flush the eye for 15 min with tepid clean water.[1] Never attempt to use a neutralizing agent in the eye. Once the eye is flushed, get the person to an ophthalmologist immediately. Permanent eye damage can occur in minutes.

To clean up a large floor or bench spill, form a barrier around the area with a special neutralizing clay designed for the purpose. Several companies sell acid spill kits (see Figure 5.1). NEVER use sawdust or any other organic barrier. Acids such as nitric acid (HNO_3) can react violently with organic matter. Fire, along with poisonous fumes, can result. The special clay should be pushed gradually into the spill. The mass should be disposed in a special waste container designed for that purpose. All solid wastes should be hauled away to a proper site by a qualified company. The remaining residue should be scrubbed with a sodium carbonate (Na_2CO_3) 5% solution and thoroughly rinsed. When cleaning a spill, make sure the skin, eyes, and lungs are protected. Small spills can be simply neutralized with sodium bicarbonate and then wiped up. Any spill greater than about 25 ml should be blocked off with the clay barrier.

In making acid reagents or solutions, workers should be aware that some acids are highly exothermic when mixed with water or other agents. Always use an ice bath when mixing solutions. Always add acid gradually and carefully to water, with continual mixing: NEVER the other way around. Acids such as sulfuric acid (H_2SO_4) will sputter violently if water is placed on top of them. When working with highly fuming acids, such as nitric (HNO_3) and hydrochloric acid (HCl), make sure adequate ventilation is provided. Be sure to wear protective gloves, apron, and goggles when handling acids (see Figure 5.2). Clean up all spills in the work area. Be sure that acids are stored in their own area in a cool, dry, dark room. Nitric acid, especially, breaks down in the presence of light. Make sure all organics and strong oxidants and reductants are kept away from acids. If acid is used to adjust pH in the plant or to strip and clean filtering media, make sure the proper resistant equipment is used. Acid products, such as Oakite®, usually come with very specific instructions as to their use. Call the manufacturer if there is any doubt as to how a product is to be stored and handled.

5.2 SOME SPECIAL PRECAUTIONS

There are some acids that require special precautions in their safe handling. The major problem with hydrochloric acid is its fuming. The TLV of HCl is 5 ppm.[2] It is primarily a pulmonary irritant. Because of this, any person rescued from overexposure to fumes should be observed in a hospital for 72 h. Like some of the other lung irritants, HCl may show delayed symptoms of lung damage.

Nitric acid has two very bad problems associated with it. Its TLV is 2 ppm.[2] When HNO_3 is exposed to light, two dangerous byproducts are produced: nitric oxide (NO) and nitrogen dioxide (NO_2). These are deadly pulmonary irritants that can cause severe pulmonary edema.

A

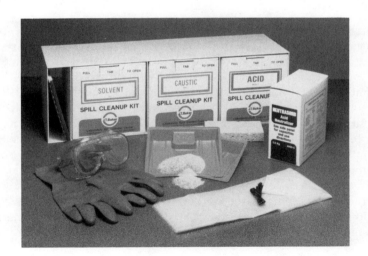

B

Figure 5.1 Proper spill control. (A) A lab should be prepared to handle large and small spills. (B) Make sure the right neutralizing agent is used for the right kind of chemical. (Photos courtesy of Lab Safety Supply, Inc., Janesville, WI.)

64 CHEMICAL HAZARDS AT WATER TREATMENT PLANTS

Figure 5.2 Proper protection. For work with strong chemicals, one must protect the eyes and skin. (Photo courtesy of Lab Safety, Inc., Janesville, WI.)

Here again, any worker injured by nitric acid should be under hospital observation for 48 to 72 h. Burns of the skin and eyes should receive particular attention. Nitric acid has a tendency to severely ulcerate the burn areas. Another problem associated with nitric acid is that it IS an oxidant. It reacts violently with organics and reducing agents. When using nitric acid for chemical analysis, one should be careful to mix reagents slowly and under a gas hood. Never store nitric acid with other reagents of any kind. Keep it out of direct light when not using it.

Ventilation is important in using both of these acids. There have been documented cases of tooth erosion, chronic bronchitis, and eye damage due to chronic low-level exposure to acid fumes. Most TLVs are set to avoid these problems. After working with acids in the field, a worker should shower and change into clean clothing. Cautious handling and good hygiene is 90% of the battle for a good safety record in your plant.

5.3 A WARNING ABOUT HYDROFLUORIC ACID AND ITS BYPRODUCTS

There is only one application for hydrofluoric acid (HF) in the water quality field. It is a reagent for a little-used gravimetric method for the determination of silica. It is also occasionally used in certain sediment metal determinations. Most of the time chemists try to avoid the use of HF because of its extreme toxicity and corrosiveness. It must be kept in special teflon containers. It corrodes metal and etches glass. It fumes excessively and requires a good gas hood for work. The TLV is 3 ppm.[2] It is reactive with organics. Once used to fluoridate water, Madison, WI was the last place to use this agent.[3] Because of its hazards, Madison has opted for safer substitutes. Lab personnel should treat HF injuries with the utmost respect. No lab should work with HF without additional chemical and first aid training in the use of this compound. If the lab must use HF, the following antidote, an 0.2% solution of alkyldimethyl benzylammonium chloride in distilled water, should always be kept near the work area.[2] This is also available under the trade name Zephiran.[4] If HF is splashed in the eyes, the area must be flushed with water immediately for at least 15 to 20 min. Plain cold compresses are then placed on the eyes, and the person is rushed to an eye specialist immediately. DO NOT ATTEMPT TO MEDICATE THIS KIND OF EYE INJURY WITHOUT THE GUIDANCE OF A TRAINED SPECIALIST. Your job is to flush and protect the eyes prior to rushing the person to the hospital. When calling 911, inform the emergency personnel that HF poisoning of the eyes has occurred. A skin injury should be flushed for 15 to 20 min with clean water. Remove and destroy all contaminated clothing. Wrap the skin burn with cold, clean cloths soaked with the Zephiran solution. Rush the person to the hospital. Because HF injures the nerves as well as the skin, further treatment is the province of an environmental physician. Very often there is severe ulceration of the injury site. These burns take a long time to heal. If every trace of HF is not cleaned off, the remaining chemical can cause further injury. If the lungs are injured, remove and resuscitate the person and rush him or her to the hospital. Both external and lung injuries require the patient to be observed in the hospital for 24 to 48 h, before being sent home.[2]

Since HF is so toxic, other substitutes for fluoridation of water supplies are used. These substitutes, in order of their preference, are sodium silico-fluoride (Na_2SiF_6); fluospar (CaF_2), which is not very common; hydrofluosilicic acid (H_2SiF_6), and sodium fluoride (NaF). These chemicals are still highly toxic and require protection of the skin, lungs, and eyes, during handling. They are somewhat corrosive, and care and attention must be given to any fluoridation equipment. Manufacturer's instructions regarding these chemicals must be followed to the letter. Dosages in drinking water must be strictly followed. Fluoride greater than 1 mg/l in water may cause pitting tooth enamel and other effects if consumed for long periods of time. To avoid errors in dosages and handling, fluoride compounds are kept in separate and locked facilities. Dosage calculations and fluoride residuals are checked frequently. The dosage area of the water treatment plant is also kept under severe security, and log-in and log-out records are kept daily. Shipments of the chemicals should be witnessed by at least two water treatment personnel. This avoids possible error in the delivery and usage of the chemicals. After working with any of these agents, the operators should shower and change into clean clothing.

Acute injuries caused by fluoride are severe and require immediate attention. However, the chronic low-level exposure to fluoride agents can be just as devastating. Once more, the symptoms of poisoning may take weeks or months to appear. The symptoms are often confusing, mimicking other sickness. Fluoride is an essential mineral needed for strong teeth and bones. However, the dosage for good health is extremely small. No more than 1 to 2 mg per man per day from all sources should be allowed.[5] When intake exceeds 2 or 3 mg/d, symptoms of chronic poisoning gradually appear. The first symptom is pitting of the tooth enamel, called fluorosis. As exposure continues and dosages increase, symptoms become more severe. The kidneys, skin, liver, heart, and CNS can also be affected. In areas with high natural fluoride levels or high levels due to poorly run fluoridation facilities, persons exposed to the substance can show a skin discoloration called a chizzola maculae. This appears as a brownish area of skin inflammation.[6] With continued exposure the teeth and bones become extremely brittle, and other symptoms of kidney and heart trouble begin to appear. Fluorine can preferentially cross the placenta, so pregnant lab workers should be especially careful in the use of fluoride compounds.[6] Any worker who has kidney problems should be doubly careful in handling fluoride compounds. Their tolerance to fluoride could be very low.

Poisoning can be avoided by stringent hygiene:

1. Always wear a respiration mask when handling dry fluoride salts.
2. Protect the hands with resistant gloves, and wear eye protection.
3. When working with any acid fluoride, such as HF or H_2SiF_6, make sure that the work area has adequate ventilation.
4. See that fluorides are stored in their own separate cool area under lock and key.
5. When dosing water supplies, have at least two people verify calculations and equipment settings.
6. When receiving shipments of fluorides, at least two people should sign in the shipments as to use and placement in the work area.
7. After using fluorides, make sure the work area is thoroughly cleaned.

8. The worker should clean himself up after the work is completed.

Such precautions can avoid tragedy later.

5.4 A WORD OF CAUTION WITH CAUSTICS

The bases could be roughly divided into two groups. The weak bases include sodium bicarbonate ($NaHCO_3$), sodium carbonate (Na_2CO_3), and calcium oxide and hydroxide [CaO and $Ca(OH)_2$)] The strong bases are sodium hydroxide ($NaOH$) and potassium hydroxide (KOH). All of these agents are used in chemical analysis and in certain field applications involving pH adjustment. They can all cause burns. In many ways the weaker bases could be more dangerous in that the injury is not immediately apparent. For example, an operator who has been carelessly handling calcium oxide or hydroxide may not discover he has been burned until several hours later. Of course, with the strong bases a soapy feel will warn the worker immediately to take corrective action.

It is important that any basic burn be dealt with quickly. Bases do not char or denature protein in the same way as acids do. Bases tend to dissolve protein. A protective scab is not formed. Thus, the burns tend to be a lot deeper and heal more slowly. The eyes, in particular, can sustain deep and lasting damage of the cornea.[1] If a base is splashed on the skin, one should flush the area with water for 15 min. A 5% solution of acetic acid or just plain old house vinegar could then be placed on the area for a minute. Then, the skin is washed another 5 to 10 min with clean water. Make sure all traces of base are removed. No soapy sensation should remain when the skin is touched. Carefully inspect the injury site for blisters. If any redness or discoloration is visible, wrap the injury in a wet, clean cloth and get the person to a burn center.

The eyes are a special problem. If a base is splashed in the eyes, it may take up to 15 to 30 min of irrigation with clean, tepid water to remove the offending substance. DO NOT delay. Just 30 sec of hesitation can result in extensive scarring of the eye tissue, with resultant blindness.[1] Once the eye is cleaned, the worker should be taken to an ophthalmologist for further care.

When working with basic substances, be sure to wear protective clothing, gloves, and goggles. Some bases, such as calcium oxide, have a severe dust problem. Thus, ventilation and dust control become very important. Spills can be contained using special neutralizing barriers and clays pushed gradually into the area. Make sure that the right adsorbant is used for the job. As with the acids, there are several excellent spill-containing products on the market. The resultant mass must be disposed of in an environmentally approved manner. The area should be scrubbed down with vinegar and water after the major part of the spill is cleaned up. Small spills of less than 25 ml can be neutralized with vinegar and then cleaned up. These simple procedures can insure a safe lab when working with bases.

There are two bases that present special problems in use and storage. These are calcium oxide and sodium hydroxide. Sodium hydroxide, or caustic soda, presents special problems in use and storage. It can either be shipped in pellet form or most commonly, as a 50% solution. Pellet NaOH must be mixed to a desired concentration. Mixing this base with water generates a large amount of heat and fumes. Provisions must be made for cooling the exothermic reaction and for ventilation. Mixing small amounts can easily be done in an ice bath under a gas hood, with continuous stirring. Whenever mixing this base in large amounts, one should be completely suited up in a respirator, goggles, and protective rubber gloves and clothing.[3] It is a good idea that after working with any base in the field to completely shower off and change into clean clothing. Even with all this protection, some traces of base can get on the skin. Store the pellet form in a cool, dry area. The base is very deliquescent, that is, it readily adsorbs moisture from the air. Store the 50% solution in special insulated tanks. This solution will begin to crystallize below 54°F or about 12°C. Resistant materials such as high quality plastics or glass are required for storage. Glass containers must be inspected frequently, as sodium hydroxide will slowly etch glass.

Calcium oxide, or quicklime, is used in the water-softening process, for enhancing coagulation, and for conditioning sludges during dewatering. It is available in 50-lb bags or in bulk shipment. It is deliquescent and will cake up on standing in a humid area. Storage areas should therefore be dry and cool. Never store coagulation salts, such as alum and ferric sulfate with lime. Calcium oxide reacts violently with these two salts. It actually seeks out the water of crystallization of these salts, with a violent evolution of heat. Thus, quicklime needs its own storage area. This agent also requires protective clothing and eye protection. Much dust is generated when slaking lime. A respirator needs to be worn while mixing it. The slaking of quicklime to calcium hydroxide generates heat. Provisions should be made for cooling the reaction vessel. Burns from exposure to lime agents may not become immediately apparent. After a period of work with lime, the worker should shower and change into clean clothing. Sometimes during showering a worker could use a 1% rinse of vinegar. This would neutralize any trace of lime. Be sure to rinse the vinegar off during showering.

Bases and most acids are not particularly toxic, but they are corrosive. They require some forethought in their storage and handling. With a few simple precautions, these agents need not cause injury. A short summary of the proper handling of the most commonly used acids and bases is provided at the end of the chapter (see Tables 5.1 and 5.2).

Table 5.1 Safe Handling of the Acids

Acid	Applications	Reactivity			Precautions and First Aid
		Water	Organic	Other	
HCl	Analysis	Mild	Mild	Nitrogen compounds, severe	Ventilation; protective clothes; flush with baking soda and water
HNO_3	Analysis	Mild	Severe	Reductants, severe	Ventilation; protective clothes; flush skin with water and baking soda; lung injuries require hospitalization for 24 to 72 h
H_2SO_4	Analysis pH adjustment	Severe	Severe		Be very careful when using water and other acids; flush with baking soda and water
H_3PO_4	Analysis, iron sequestering agent	Severe	Severe		Be careful with water and other chemicals; flush burn with baking soda and water; highly corrosive to all materials except high-quality teflon
HF H_2SiF_6	Analysis	Mild	Severe	Glass	Highly toxic; can form severe ulcerated burns; flush burn with water 20 min; apply zephiran solution; rush to hospital

Table 5.1 (continued)

Acid	Applications	Reactivity			Precautions and First Aid
		Water	Organic	Other	
Salts of flouride CaF_2, NaF Na_2SiF_6	Analysis; flouridation of water supplies				Highly deliquescent; dust hazard; corrodes glass and steel equipment; keep under lock and key; highly toxic >3 mg/d; can burn skin; use same first aid as with HF; wear complete protective clothing with ALL F compounds; forced ventilation

Table 5.2 Safe Handling of the Bases

Base	Application	Reactivity	Precautions and First Aid
Alkali earth bases NaOH, KOH	pH adjustment and reagents	Violent reaction with water; mix in ice bath under a gas hood	Flush burns with water, then vinegar, then water; go to burn center; ventilation and protective clothes
$NaHCO_3$	Reagents, first aid for acid burns	Mild base of little hazard	Wash thoroughly after handling
Na_2CO_3	Reagents, pH adjustment and softening	Evolves heat with water; dust hazard	Ventilation and protective clothing; wash with water 15 min
Limes CaO, $Ca(OH)_2$	Softening	Deliquescent; evolves heat with water; violently reactive with coagulent salts; keep in separate storage facilities	Ventilation and protective clothing; flush skin with water 15 min

REFERENCES

1. Steere, N.V., Ed. 1971. *CRC Handbook of Laboratory Safety, 2nd ed.* CRC Press, Boca Raton, FL.
2. Proctor, N.H., and J.P. Hughes. 1978. *Chemical Hazards of the Workplace.* J. B. Lippencott, Philadelphia.
3. AWWA Inc. 1971. *Water Quality and Treatment: a Handbook of Public Water Supplies,* 3rd ed. McGraw-Hill, New York.
4. 1983. *Merck Index* ,10th ed. Merck, Rahway, NJ.
5. Kirschmann, J.D. 1975. *Nutrition Almanac.* McGraw-Hill, New York.
6. Waldbott, G.L. 1978. *Health Effects of Environmental Pollutants.* C.V. Mosby, St. Louis.
7. Cameron, D., and G.A. Hartson. 1988. Aluminium and flouride in the water supply and their removal for haemodialysis, *Sci. Total Environ.* 76:19–28.

6
Dangers of Strong Oxidants and Reduction Agents

6.1 DANGERS IN THE PLANT

In the water and wastewater treatment plant, there are five or six agents that are frequently used that can become real fire hazards if handled incorrectly or stored improperly. Some agents are used for disinfection. Others are used for iron removal and taste and odor control. Rarely anticorrosion agents may be needed. Other agents, such as lubricating oil, while not harmful in themselves, cannot be mixed or stored with certain oxidants without danger of combustion or explosion. The operator should be aware of the incompatibilities of the chemicals in his or her work area.

These reactive chemicals are oxidants and reductants. An oxidant is an agent that will take electrons from another reactant. Thus, the oxidant gains electrons, and the substance acted upon by the oxidant loses electrons. A reducing chemical is a substance that gives or donates electrons to its reactant. Thus, the reducer loses electrons and its partner gains electrons. Some reactions involving a change in valence state are slow, but other reactions can proceed very rapidly, with violent evolution of heat.

Potassium permanganate ($KMnO_4$) is used for iron and manganese removal as well as for taste and odor control. In its granular form it is easy to store and apply. However, its one serious drawback is that it is absolutely incompatible with any organic material. If some of the dry salt is even spilled on plain paper, it could spontaneously ignite. Thus, it must be stored in its own cool, dry area. It must also be stored and delivered in corrosion-proof equipment. All spills must be thoroughly cleaned. The work area must be completely hosed down after handling permanganate. Any rags that have been in contact with the salt or solution should be destroyed or disposed of in a fireproof container. Attention to cleanliness can avoid accidents in working with potassium permanganate.

One would not think that the activated carbon used for taste and odor removal would be a potential hazard. Yet under certain conditions, activated carbon can also be dangerous. Normally, there is no dust hazard with the grades of carbon used in treatment. It WILL, however, react with such materials as calcium hypochlorite [$Ca(OCl_2 \cdot 4H_2O$)], sodium chlorite ($NaClO_2$), potassium permanganate ($KMnO_4$), and oils.[1] Thus, this substance also needs its own cool, dry storage area. After working with carbon one must make sure the work area is thoroughly cleaned. If a smoldering carbon fire does occur, do not blast the area with a heavy stream of water. This scatters sparks.[1] Use a fine foam or water mist. If possible, isolate the smoldering material in a steel container and haul it out of the building to extinguish it.

Only rarely are the sodium chromate salts, such as sodium dichromate ($Na_2Cr_2O_7$), used in water treatment. The chromate salts are usually used as corrosion-control agents in cooling or heat-exchange water. Sometimes they may be used in their acid form as cleaners. They are highly toxic and must be handled as any heavy metal poison. They will also react vigorously with organics and any reducing chemical. They can corrode metal severely. Thus, these agents require storage in resistant containers rated for corrosive substances. They demand separate facilities for storage and handling.

If one wants real trouble, one can usually find it in the hypochlorites and chlorites. Although smaller plants prefer these agents for disinfection, they can present problems. They are incompatible with just about anything. Sodium chlorite should never be stored anywhere near coagulant salts such as alum. These agents react violently, producing a toxic gas. All the hypochlorites and chlorites are reactive with petroleum products. With explosive force these salts will generate a mustard-like gas if mixed with oil or gasoline. The byproducts can cause rapid and severe lung edema. It is vital that these salts be stored in resistant containers in a cool, dry place. They should only be used in those steps of water treatment that pose no reactive danger with other chemical agents. For instance, alum treatment should be done first, and then the disinfection process should take place. Or if prechlorination must be done, make sure that process is done far enough away from coagulation as to not pose a danger. Any exposure to an oxidant fire should be treated as a chemical burn and a lung injury. The injured party should remain in the hospital for at least 72 h to assure that lung injury is properly treated. Be a stickler for cleanliness and safety with these agents.

It can be seen that there are many areas in the field that can be potential fire hazards. In addition to the worker's efforts in running a safe and clean establishment, management should initiate safety programs for their plants. Every treatment plant should form a close alliance with their local fire department. The fire department should be informed of the kinds and locations of all potentially combustible chemicals. This makes for a more effective and safer response from the fire department, should an emergency occur. It also protects the firemen from potential toxic injury. The plant itself should have the right kinds of fire extinguishers at frequent and conspicuous stations throughout the work area. First aid kits and fire blankets should also be kept in conspicuous sites throughout the plant. Every year the plant should have a complete fire inspection, and identified problems should be corrected.

6.2 DANGERS IN THE LABORATORY

In many ways the lab can be a far more dangerous area than the shop or treatment area. Many kinds of strong oxidants and reductants are used in the course of daily analyses. Some of these agents can form explosive byproducts upon sitting and aging. Some processes used in analysis are violently exothermic and must be handled with great care. There are five common agents that a lab technician needs to handle with care. These are nitric acid (HNO_3), perchloric acid ($HClO_4$), the ether peroxides, the chromate and dichromate salts (CrO_4^- and $Cr_2O_7^=$), and potassium permanganate. This is by no means a complete list. Rather, the author has chosen to point out the dangers of those agents most commonly used in most water quality labs. Each lab may have additional hazards that need to be addressed. It is the responsibility of the lab supervisor to research each reagent as to its potential explosion and fire danger.

In addition to instituting safe practices, the supervisor must also give special attention to the proper storage of each class of chemicals. There are several excellent references that class chemicals as to their reactivity and compatibility with various materials. These references, in addition to the MSDS sheets shipped with each chemical, should be kept in accessible areas of the lab. As in the field, the lab should work closely with the local fire department in instituting a sound safety program. Storage areas should be cleaned out and inventoried at least once a year. Close attention needs to be given to classing chemicals according to compatability. For instance, it is safe to store agents such as ascorbic acid, barium chloride, and sodium chloride together. But agents such as ethyl ether, perchloric acid, or certain organics need their own separate, refrigerated storage areas. Refrigeration facilities must be explosion proof. Again, the proper fire extinguishing equipment must be highly visible and handy and kept in good repair. First aid kits and fire blankets must be kept up to date.

Before reviewing how certain chemicals should be handled, the issue of first aid for burns needs to be addressed. Burns are the most serious of all injuries, for a number of reasons. The skin is a major protective defense for the body. Even in sweating, the skin, in a way, serves as a third kidney. A significant amount of waste is excreted through sweat. The skin is a barrier both to microorganisms and many chemicals. The skin is a cooling device and a electrolytic barrier. The body is covered with sweat glands, and they are a significant factor in cooling. A person born without sweat glands, due to a genetic defect, will have much trouble in maintaining proper heat balance in the body. A major worry with burn victims is that extensive areas of injury allow the loss of body salts and fluids. Many burn victims die of electrolytic shock or infection. Thus, it is paramount to first stop the fire in the workplace and remove the victim from the area of danger as quickly as possible. Next, the burns must be gently cleaned and protected with cold, wet cloths, and the person rushed to the hospital. Good first aid is critical to recovery.

In a lot of violent reactions, the person's clothes may catch fire. If the victim does not grab a fire blanket and roll to smother the fire, he should be forcibly knocked down and rolled in a fire blanket to smother the flames. Even without a fire blanket available, the drop and roll technique is the best way to smother a fire. DO NOT RUN.[2] This only fans the flames.

Once the fire is put out, apply a gentle stream of cool clean water to the area. If there are only first degree burns, the immediate application of clean, cold water may prevent further blisters. If possible, gently pull charred clothing away from the burn and continue to apply clean water. If the injury looks severe, leave the removal of clothing to the hospital personnel. DO NOT attempt to debride (that is, remove damaged clothing and skin) third degree burns. This leads to the question, how does one identify the seriousness of an injury? The degree of injury has these criteria:[3]

1. First degree — only redness of the upper skin, slight warmth to the touch, pain, and maybe some blisters. Cool water or ice on the area may be the only medicine needed.
2. Second degree — blisters definitely form. The area appears red, angry, and raw. Much pain. Clean gently with clean, cool water. Wrap in clean, wet cloths. Call 911 to rush to a hospital. Some permanent scars form.
3. Third degree — very serious. Pain is not felt, because nerves are

destroyed. Charred, white, lifeless-looking areas down to the muscle. All three layers of the skin are damaged. Severe scarring occurs, with the complete destruction of sweat glands and hair follicles. Clean, wet cloths should be wrapped on the burn. Call 911 to rush to a hospital.

In all cases, do not apply butter, oil, or any type of cream or ointment. This may only encourage infection. Your job is to clean and protect the area and to get the person to a medical expert.

Having addressed the human side of a laboratory fire, attention will now be given to the proper containment of a fire. Because things happen quickly in a fire, any potentially dangerous experiment or analysis should only be undertaken on the buddy system. It is doubtful that only one person could contain a lab fire. First, all labs should be provided with smoke alarms and a sprinkler system. The alarm should be piercing enough to be heard all over the building and should be wired into the local fire station. The right fire extinguishers for the right kind of fire should be placed in conspicuous sites all over the lab. In the author's lab area, there are a total of five extinguishers for a relatively small area. The ratio of extinguishers should be at least one for each room. In addition, there should be fire blankets supplied in dangerous areas of the lab. For example, a station should be by the atomic adsorption area, and another by the wet digestion area of the lab. First aid kits and fresh, clean cloths should be maintained in various areas throughout the building (Figure 6.1 illustrates a complete safety station). A well-provisioned lab is a safe lab.

It is well known that different fires require different kinds of extinguishers. Electrical fires or certain organic fires simply cannot be put out by water and soda extinguishers. They require either a carbon dioxide- or a freon-based (halo-carbon) type of extinguisher. Thus, it is important that personnel be trained in types and uses of various extinguishers. Here is a description of four basic kinds of extinguishers (Figures 6.2 and 6.3):[2]

1. Water or water-soda — This is simply plain water, or a mix of water and soda under pressure. A lab can also have water outlets supplied with fire hoses. These extinguishers are good for simple paper or inorganic oxidant fires. DO NOT use them on electrical, organic, or oil fires.
2. Carbon dioxide — This is simply a canister of compressed carbon dioxide and is very good for electrical fires and organic fires. CAUTION: carbon dioxide is extremely cold when suddenly released from pressure. One can get severe ice burns from carbon dioxide.

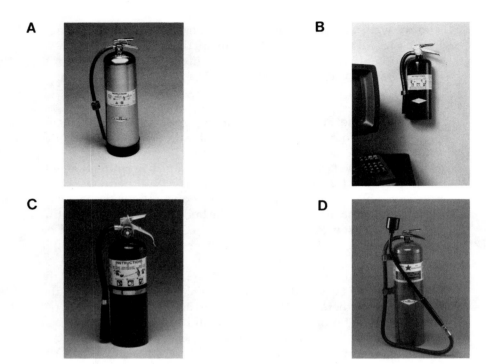

Figure 6.1 Types of fire extinguishers. Make sure the right extinguisher for the right fire is used: (A) water; (B) dry chemical, such as carbon dioxide; (C) Halon electrical; and (D) metal fire. (Photo courtesy of Lab Safety Supply, Inc., Janesville, WI.)

3. Dry Chemical — These contain sodium or potassium bicarbonate. They are used for certain flammable liquid fires that cannot be controlled by carbon dioxide.
4. Halo-carbon extinguishers — These contain a freon-like material to exclude air and smother a fire. They are used on electrical fires and certain metallic fires. Be careful; freon-like gases collect in low-lying areas and can asphyxiate a worker.

Make sure the worker knows which fires can be put out with which extinguishers.

It would be wise to work with the local fire department in setting up a sound safety program. First, the fire department should be aware of just what type of chemicals are handled by the water quality lab. This prepares them for emergencies that might be encountered in fighting a lab fire. It also protects the firemen from poisoning. A fire inspection should be performed on the lab at least once a year. Suggestions should

DANGERS OF STRONG OXIDANTS 79

COMPARISON OF FIRE EXTINGUISHER TYPES

Type	Advantages	Disadvantages	Notes*
HALON CLASS A, B, C or B, C	• Quick Fire Knockdown • Will Reach Hidden Fires • No Damage To Equipment • Good Visibility • Good Discharge Range • Heat Absorber	• Requires Rapid Discharge • More Expensive • Personnel Hazard (Halon 1211) • Not For Deep-Seated Fires	• Most Common System For Electrical/Electronics • Maximum Effectiveness Requires Rapid Detection
DRY CHEMICAL CLASS A, B, C	• Good On Oil/Grease • Good Knockdown • Low Cost	• Limited Personnel Hazard • Equipment Damage Likely • Cleanup Required • Not Suitable For Hidden Fires	• Compatible With Other Agents • Subject To Equipment Interference
CARBON DIOXIDE CLASS B, C	• Good Fire Suppression & Cooling Capability • Will Reach Hidden Fires • No Equipment Damage • No Messy Cleanup • No Odor	• May Be Toxic To Personnel • May Cause Thermal/Static (Shock) Damage • Heavy Vapor Settles Out Limiting Total Discharge Range	• Secondary Choice To Halon When Fighting Class B & C Fires
SODIUM BICARBONATE CLASS B, C	• Won't Bake On • Easy Cleanup • Good Knockdown • No Odor • Non-Conductive	• Not Suitable For Hidden Fires • Slight Respiratory Hazard	• Secondary Choice To Halon When Fighting Class B & C Fires

Figure 6.2 Summary of classes of fires. Many kinds of fires can occur in a chemistry lab. Place various classes of extinguishers at strategic places within the lab. (Chart with permission of Lab Safety Supply, Inc., Janesville, WI.)

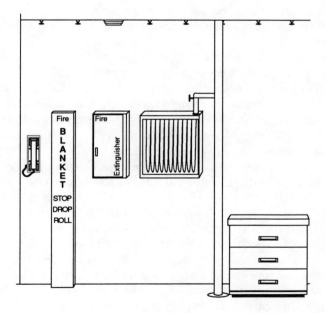

Figure 6.3 A fire station for a hazardous area. Each area should be evaluated as to the types of hazards presented. Appropriate fire equipment should then be provided.

If a fire flares up, the first concern, after rescuing personnel, should be containing the danger. The *CRC Handbook of Lab Safety* recommends that action against a fire should include these steps:[2]

1. Alert other personnel to flee the area.
2. Contain the danger. Small bench fires can easily be put out by a small extinguisher, but small fires have a propensity for escalating into large disasters. If possible, close off the room of the fire.
3. While containing the fire, two other things must happen at the same time: evacuate the building and summon the fire department. Most modern buildings have smoke alarms that are directly linked with the fire station. If there is no automatic alarm link, make sure that someone calls the fire department. Due to the toxic nature of even small bench fires, evacuation should take place, no matter how small the fire.
4. Rescue any injured parties and initiate first aid procedures.
5. If the fire is within limits of your control, contain the emergency and begin extinguishing procedures until the arrival of the fire department.

Make sure that fire drills are held at regular intervals so that responses to an emergency become automatic. As with gas leak emergencies, workers should also be trained in the use of self-contained breathing equipment.

Now that general points of fire emergency have been covered, one can deal with the dangers of specific chemical agents. Nitric acid, in addition to being corrosive, is a very strong oxidant. Make sure that it has its own cool, dry acid cabinet away from other acids. Follow directions in its use, to the letter. If necessary, supply a cooling bath for reactions. NEVER try to contain a spill with sawdust or any other organic adsorbant. Only use those barriers that are rated inert to strong oxidants. After containing a spill, clean up all traces of the nitric acid. Remember, the nitrite and nitrate salts are also strong oxidants. Do not store them with other organics or reductants. Handle such salts with care, and thoroughly clean up the area after the use of these salts.

Strong dichromate, chromate, and potassium permanganate salts are strong oxidants that can cause a fire when contacted with paper, cloth, or other organics. Reductants such as hydroxylamine hydrochloride are also strong reactors with certain oxidants. Strong reductants should be kept separate from strong oxidants. When proceeding with the use of strong oxidants or reductants, one should be alert for the evolution of heat and gas. If one is not sure of the vigor of a reaction, one should mix reagents in an ice bath and under a hood. Make sure that such chemicals are stored in inert glass or resistant teflon containers. The storage area should be cool, dry, and dark. Some of these agents break down in the presence of light. Inventory and discard those chemicals whose shelf-life may have expired. Make sure that disposal is carried out in a proper manner.

Although the hazards of ethers have been covered before, their tendency to generate peroxides makes it important that the chemist is aware of potential explosion dangers. Ethers have two problems that must be considered: they have extremely low flash points, and many ethers can generate peroxides, upon aging. The peroxides can endow the ether with explosive bomb-like properties. Even the friction of unscrewing a cap can set off an old can of ether. And a fresh can of ether can catch fire after contact with even static electricity from clothing! Consider that the flash point of ethyl ether is only –45°C. Its hazardous fuel-to-oxidant ratio is about 2 to 48%.[2] Thus, working with ether under even the best of conditions, requires a good ventilation system and special provisions for preventing sparks and open flames. It is paramount that proper handling and storage of ethers be practiced.

The author prefers that a can of ether be kept only 4 to 6 months after breaking the seal. Instances have been recorded of peroxides forming in ether in as little as 2 months of use.[2] It is also recommended that ether be stored in dark glass containers and in an explosion-proof refrigerator. Glass seems to discourage the formation of peroxides. A chemist can test for the presence of peroxide in ether by using a potassium iodide test (10% KI) or a mix of ferrous ammonium sulfate 1%, 1 N sulfuric acid, and 0.1 N ammonium thiocyanate. A little of the testing agent is gently mixed with a little ether. The presence of a yellow-to-brownish color means the presence of peroxide.[2] Get a qualified agency to properly dispose of the ether. If you do not know the age of the ether, or if there are crystals or a oily liquid in the bottom of a bottle of ether run, do not walk, to call the local fire department. They will get in contact with a bomb squad. The bomb squad will properly transport and destroy the old ether. The job of the chemist is to block off the area in which the old ether was found and to prevent people from going into the area until the ether is safely removed. Many ethers can form peroxides. Ethyl ether that has not been stabilized is the worst offender in this respect. This is the most frequently used ether in the water quality field. Whatever ether is used, the chemist should look up the properties of that agent and take pains to handle it correctly. The *CRC Book of Lab Safety* has an excellent section about the dangers of ethers in the lab and gives detailed directions as to their safe handling. Every lab should have this book on hand.

As was stated before, perchloric acid ($HClO_4$) is indispensable for the digestion of sediment samples for phosphate and heavy metal tests. It is sold at 70 to 72% strength by weight. Although it looks like an ordinary acid, it has the rather extraordinary property of reacting like a strong oxidizer and dehydrator at elevated temperatures. It violently attacks organic material and other chemical reductants. The degree to which it reacts can be said to be nothing short of explosive. This acid requires the utmost caution and respect. Because even the residues of this acid are explosive, a special wash-down gas hood is used when working with this agent. If a corrosion-proof wash-down hood is not in the lab, one should not use perchloric acid for any reason. The peculiar properties of perchloric acid require rigid adherence to safety rules:

1. This is a strong acid and will severely burn eyes, lungs, and skin. Use ventilation and protect eyes and skin when working with this acid.
2. Hot, concentrated solutions, such as those used in sediment digestions, are strong and explosive oxidants. Make sure that they are used under a wash-down hood that also has a corrosion-proof bench. Materials used in making the hood and bench should also be impervious and easily cleaned. A highly resistant stone or high-grade stainless steel can be used. Seams that join parts of the bench and hood should be resistant to leaks and seepage. Floors around the perchloric acid station should be sloped to a drainage area and made of concrete that is sealed with a highly resistant epoxy paint. Check the floor at frequent intervals for wear and spills. Spills should be cleaned thoroughly, and the floor repaired.
3. Metallic residues of perchlorate and anhydrous perchloric acid can be detonated just by touch or friction. If magnesium perchlorate ($Mg(ClO_4)$) is used as a desiccant, make sure that the spent salt is disposed of with copious amounts of water. Be meticulous in cleaning up spills of either acid or salt. At least three thorough washes of a work area should be done.
4. Perchloric acid has a violent and nasty tendency to attack any reductant or organic chemical. Even when used for digestions, the acid should be added to the reaction slowly and in small steps. Never deviate from the directions given for a particular chemical analysis. GO SLOWLY. Do not hurry a perchloric acid digestion.
5. Because the acid is incompatible with just about everything, it should be kept in its own explosion-proof refrigerator. Only a high-quality borosilicate glass, or resistant teflon should be used to contain the acid. Never keep more than 14 lb (about 1 gallon or maybe 3 or 4 l) in the entire lab at any one time. Discard any acid that is beyond its recommended shelf life.

It is highly recommended that every lab handling perchloric acid keep the *CRC Handbook of Lab Safety* on hand. The Association of Official

Agricultural Chemists and the Factory Mutual Engineering Division also have excellent references concerning the proper handling of perchloric acid.²

When working with any strong oxidant or reductant, the worker should be aware that, under certain conditions, these chemicals have bomb-like properties. A well-organized safety training program and a rigorous ground maintainance program can go far in preventing tragedy.

REFERENCES

1. AWWA Inc. 1971. *Water Quality and Treatment: a Handbook of Public Water Supplies*. McGraw-Hill, New York.
2. Steere, N.V., Ed. 1971. *CRC Handbook of Laboratory Safety*, 2nd ed. CRC Press, Boca Raton, FL.
3. Holvey, D.N., and J.H. Talbott, Eds. 1972. *The Merck Manual of Diagnosis and Therapy*, 12th ed. Merck, Rahway, NJ.

7
Microbial Hazards in Water

The dangers of exposure to pathological agents are primarily a concern of those who work with municipal and industrial wastewater. There are many opportunities for a worker to become inadvertently infected with various microorganisms. Some sicknesses are mild, such as the astrovirus stomach upsets. Others can be quite severe. The threats of typhoid, tetanus, hepatitis, and polio are still very real.

There are three very basic modes of transmission for the enteric and waterborne agents. The air route is primarily a route for viruses, but some eggs of certain parasites can also be carried by air currents. Other agents are true enteric agents, that is, these diseases are spread through contact with human excrement. Direct hand-to-mouth contact is the pathway of entry. This is not only a problem with sewage treatment operators, but also a problem with those who handle food without thoroughly washing the hands. Then there is the pathway of mechanical exposure. Flies, mosquitoes, roaches, mice, rats, and other vermin are quite capable of spreading sickness throughout the plant. Often these pests are drawn to foods and certain chemicals. So it behooves the groundskeepers to keep a sharp lookout for unwanted "guests" in the treatment plant.

There are five broad types of organisms that can cause disease and are commonly found in the sewage treatment process.[1] These are the viruses, bacteria, protozoans, helminths (lower worm parasites), and some fungi. Each of these classes will be covered in some detail. The key to good health is prevention through an intimate knowledge of the microorganisms. One cannot avoid disease if one does not understand the biology of the agent in question.

7.1 THE VIRUSES

There are over 300 species and/or serotypes of viruses found in water and wastewater. Some are merely annoyances, causing stomach upsets and colds. Others are quite deadly, such as hepatitis A and polios I, II, and III. All, by repeated infection, can pull down the health of the water quality worker. Most can be controlled by paying strict attention to simple hygienic practices.

Since there are so many different viruses, only the most important groups will be reviewed here. Viruses are spread by all three mechanisms of exposure, that is, by hand, pest vectors, and aerosols. Polio is an example of a group of viruses that can be spread by the fecal and mechanical route. Flies, in particular, can transmit polio.[2] Thus, vermin control becomes very important. Contrary to popular belief, the three types of polio are still active. Polio particles are capable of laying dormant in soil or sludge, for years. Unlike smallpox, polio does not need a live vector to remain viable. It is always present in certain soils, and there can also be healthy asymptomatic carriers of the disease.[3] In certain third world countries it is a very real and immediate threat.

Hepatitis is strictly an enteric virus. It is spread by the fecal-oral route. This virus is highly resistant to chlorination. In 1953 New Delhi, India, experienced a failure in the sand filters used for cleaning drinking water. Despite the fact that extra chlorine was applied, the WHO estimated that fully 20% of that population was exposed to hepatitis. Chlorine levels were as high as 10 ppm. Since the aftermath of this disease includes a possibility of liver cancer later on, it is imperative that the sewage treatment operator take steps to avoid exposure. There is no vaccine for this type of hepatitis.

AIDS, which is an HIV virus, is an immunosuppressor. It is not considered a threat to water quality personnel. The conditions in wastewater are not conducive to its long-term survival. Also, this virus is sensitive to chlorination. As little as 1 ppm of chlorine for 15 min of contact time can inactivate the virus. Still, it is important for the sewage worker to protect cuts and bruises with rubber gloves while performing dirty tasks around the plant.

There are three broad classes of what the author calls the lesser viruses. Norwalk and astroviruses cause stomach upsets. These viruses are not only common in sewage, but may also be found in home wells that are too close to septic drainage fields.[4] Quite often, new sewage workers complain of frequent diarrhea during the first year of work. Gradually the workers may build up a resistance and experience good health thereafter. This is not a reason for complacency. Exposed workers can still become silent carriers of such viruses to their families. Norwalk and hepatitis viruses, particularly, can be carried without symptoms by the worker to his or her family.[5] The third kind of virus causes colds and upper respiratory problems. These are called the adenoviruses. While not serious in themselves, the adenoviruses can pull down a worker's resistance to bacterial infections. Because most viruses are very small and light, they can be carried aloft by air currents. Activated sludge tanks, sludge drying facilities, and biological filtration areas are prime sites for virus exposure.

7.2   BACTERIA

Over 100 kinds of bacteria, plus a number of different serotypes of the most common species, are found in natural and wastewaters. The most dangerous of these organisms are *Salmonella typhi* and *Vibrio cholerae*. These are the agents of typhoid and cholera. They are most numerous in primary treatment tanks, but can also be found in sludge-handling areas. Scrupulous cleanliness is the best defense against these diseases, as they are spread by the hand-to-mouth mode. Legionella bacteria are another problem. They are found primarily in aerosols around activated sludge tanks. *Legionella pneumophila* is found mainly in warm-water habitats that are high in nutrients. It appears to be resistant to chlorination, even though it is Gram negative.[6] Water-cooled air conditioners and lab humidifiers can also harbor this deadly bacteria. It behooves both the chemist and operator to make sure all air conditioning systems in the plant or lab are kept immaculately clean. In addition to these three main threats, there are at least 15 different serotypes of *Escherichia coli*, some of which are pathogenic. Most *E. coli* are harmless symbiants that live in the human intestines. But there are a few types that can cause severe stomach upset. In addition to *S. typhi*, there are at least 11 other types of Salmonella bacteria.[7] These cause varying degrees of gastrointestinal distress. There is one bacteria, *Leptospirosis icterohaemorrhagiae*, that is spread mechanically by rats and mice. It causes a severe fever with jaundice. Thus, rodent control is important in the sewage treatment plant.

Although tetanus is not considered a waterborne bacteria, it is of concern to the groundskeeper and sewage treatment operator. The agent of contamination is *Clostridium teteni*. It is a Gram-positive bacteria that is capable of forming resistant spores. It is found in soil and in the intestinal tract of cows and horses. Deep, ragged wounds are the perfect place for the growth of this bacteria. It secretes the powerful exotoxins that cause the dreaded "lockjaw." Symptoms appear 1 to 3 weeks after exposure. This disease is almost always fatal to humans. Thus, it is important that all injuries get immediate attention. The wound should be completely flushed out with clean water and then with a 2% solution of hydrogen peroxide. Stop the bleeding and bandage the wound. Deep wounds need hospital attention. If the worker has not kept up with his or her tetanus shots (every 5 years for boosters), the person should receive a tetanus shot. Many plants require proof of vaccination before a worker is hired. Vigilance is the best prevention of this disease.

Only a few of the more important bacteria have been discussed. However, the water quality worker should be aware that sewage is almost the perfect growth medium for many kinds of disease agents. It would not hurt for supervisors to have a basic training class in microbiology once or twice a year. Always the worker should be aware of the microbiology of his plant. Care should be taken to avoid the contamination of foods, and a change of clean clothing should be available for the worker before returning home.

7.3 THE PROTOZOA

There are only four kinds of protozoa that sewage workers need to worry about. There are two Amoebida, *Entamoeba histolytica* and *Giardida lamblia*; one Ciliate called *Balantidium coli*; and one Coccidida in the *Cryptosporidium* genus.[1,8] All of these are capable of forming resistant spores. These spores are likely to be found in the sludge-handling facilities. It may be possible that airborne dust can carry the spores of such organisms. All of these organisms can cause severe diarrhea and dehydration. But the worst disease agent is *Entamoeba histolytica*. This agent has been known to cause liver and even brain abscesses.[1,8] This is the famous amoeba of travelers, which is nicknamed "Montezuma's revenge." However, all of these agents can also cause a careless traveler distress. Good hygienic practices are the best defense against these pathogenic protozoa.

7.4 FUNGI

Because of the nature of sewage treatment, fungi are abundant in waste water. Trickling filters and roughing towers have fungi as a major component of their biological communities.[9] *Aspergillus fumigatus* is a mold that inhabits sludge drying beds.[1] In all, there are five or six common molds and yeasts found in sewage and even in natural clean water environments. Swimmer's ear is an infection of *Aspergillus niger*, *A. flavus*, or *A. fumigatus*.[3] These are found in almost any kind of tropical water environment and in most temperate waters during the summer.

In addition to ear and skin infections, these fungi can also cause respiratory problems. In fact, almost any area of the body can be affected by a fungal infection. Spores of the fungi can easily become airborne. That's the bad news. The good news is that a healthy worker has a pretty good resistance to most fungi. Indeed, a worker who shows repeated infections by fungi may be suspected of having a lowered resistance due to overall poor health. The immune system should be checked for abnormalities. One of the first signs of AIDS and other immune problems is susceptibility to fungal infections.

7.5 HELMINTHS AND OTHER HIGHER PARASITES

Dried sludge that is not heat treated can have, in addition to the bacteria, viruses, fungi, and protozoa, a number of higher parasites. The most common kinds include parasitic nematodes and cestodes, or tapeworms. Parasitic infection should be a particular worry of the tropical sewage worker, since there is no cold season to kill off the eggs of these pests. In fact, some sewage workers that are in good health otherwise, can carry a type of roundworm called *Ascaris lumboricordes*, without symptoms.[1] A weakened individual could show serious sickness. Just because some parasites may be endured without symptoms is no reason to become complacent. Hookworm and many kinds of tapeworms can completely disable a person. All of these diseases can be spread by the oral-fecal route (hand to mouth). Workers should be encouraged to scrub, even under the fingernails, before eating lunch. All of these parasites cause digestive disturbances and anemia. Because of the ineffectiveness of even heat treating in killing the eggs of some types of worms, it is not recommended to use treated sludges for vegetable gardens. Untreated sludge should not be used for any agricultural purpose.

7.6 PRECAUTIONS

Needless to say, the water pollution control worker is exposed to a wide variety of biological hazards. The first and foremost defense against work-related sickness is immunization. All workers should be current with their polio vaccinations. Repeated polio boosters are no longer recommended, but certainly the health record should show that the worker has received the types I, II, and III vaccines at sometime in his or her life. There are now tetanus vaccines that remain active for 10 years. Routine tetanus vaccines should be received every 10 years. However, if one is seriously wounded at work 5 or more years after the last shot, it might be wise to ask the doctor if a booster is needed. Weakened workers or those elderly workers over 65 years of age should receive flu and pneumonia vaccines. If an adult has never received measles, mumps, or rubella immunizations, it might be wise to speak to the physician about this. Certainly, every pollution worker should receive a yearly physical and keep careful record of all immunizations for himself and his family.

There are some agents that are not recommended for routine vaccination, unless the worker is going to travel and consult in a third world country. These are hepatitis A immunoglobulin, hepatitis B vaccine, typhoid, and cholera. These agents are not given routinely because hygiene is good in the United States. Also, there is a far greater chance of an adverse allergic reaction to these agents. Their use is reserved for only exceptional circumstances.

Since there are relatively few agents that are combatted by routine vaccination, hygiene becomes very important. All sewage and water treatment plants should provide their workers with a separate lunch room. This room should be kept meticulously clean. Ample kitchen facilities should include a refrigerator, a large sink, and enough counter and table space for the proper preparation of food. Locker rooms and lavatories should also have shower and changing facilities. Workers should be encouraged to scrub their hands well with a strong disinfectant soap before eating lunch. Chemists should be provided with lab coats, and field workers with uniforms. It is recommended that field workers shower and change into street clothes before going home for the day. It is quite possible for a worker to carry a waterborne disease home to the family. If either a chemist or worker is going to do a particularly dirty job, protective gloves should be worn. If any water quality worker has open cuts or any sores, special care should be taken to protect the wound from exposure. All workers should keep an extra change of clean clothes at the plant, in case of emergencies. For jobs that are going to kick up large amounts of dust or mist, the worker should wear a protective face filter. A little foresight can dramatically reduce the number of sick days caused by enteric agents.

7.7 SPECIAL PRECAUTIONS

In both drinking water and wastewater treatment, there may be times that a commercial diver is needed to repair an inlet or a piece of submerged equipment. Needless to say, divers are intimately exposed to disease agents. Common problems facing the environmental diver are eye and ear infections, skin infections, and gastrointestinal upsets.[6] Thus, the needs of these people are special.

The U.S. Navy recommends that any commercial diver that is going to be exposed to potential wastewater dangers has all the required immunizations of the wastewater treatment operator, plus typhoid, diphtheria, and (if needed) cholera shots. In addition to this requirement, these additional restrictions should be practiced:

1. Any environmental diver who has open wounds should be forbidden to dive.

2. Before and after each dive, the ears should be flushed with a 2% acetic acid and aluminum acetate solution (Domeboro's Solution).
3. In severely polluted areas a full drysuit with complete diving helmet should be used.
4. For drinking water inlets it may be acceptable to use the traditional wetsuit, face mask, and regulators. Make sure that the wetsuit is of good quality material and has a protective hood. The face mask should fit securely around the face.
5. After the dive the entire rig must be hosed down with clean water. All diving equipment should be carefully cleaned with a disinfectant solution, such as betadine. Some wetsuits are made with material that washes very well with a good detergent in a washing machine.[6]

In addition to the usual enteric organisms, the diver is often exposed to some rather bizarre organisms that are natural to the particular water environment. These organisms, while not usually disease agents, are what can be called opportunistic feeders, that is, they can cause infection in weakened individuals or, once gaining entry in a wound, can cause severe tissue damage. Species belonging to the genera of *Pseudomonas*, *Klebsiella*, and *Aeromonas* are most commonly associated with severe Gram-negative infections.[6] Many of these organisms are highly resistant to antibiotic treatment. So it behooves the diver to be well informed about the area he is assigned.

There is one protozoan that merits special mention. *Naegleria fowleri* can cause a rapidly fulminating form of encephalitis called primary amoebic encephalitis.[6] Only if proper diagnosis and agressive antibiotic treatment is given in the first 24 to 48 h can a person even hope to survive this horrible disease. Its fatal course is run in 10 days, with no treatment. Normally this amoeba lives peacefully on the bottom sediments of tropical and semitropical lakes, grazing on bacteria and fungi. However, it is an opportunistic feeder of the worst kind. It enters through the nose of a swimmer or careless diver. From there it travels to the exposed nerves of the olfactory system and then directly to the brain. Positively NO diving is to be done in organically rich tropical waters, without the protection of a full face mask and drysuit. Certainly, the local hospital should be informed if a dive is to take place in these kinds of waters, so that if some sickness does occur, proper diagnosis can be made. Misdiagnosis is frequent, since most doctors have never seen a case of amoebic encephalitis.

Environmental diving is a highly specialized field demanding thorough training. It is not an endeavor to be taken up by the operator who is a sport diver. Thus, it is wise to hire out difficult repairs to commercial companies specializing in those activities.

After reading such a section in microbiology, the water quality worker may come to the conclusion that all microorganisms are vermin. Nothing could be further from the truth. Only about 2 to 3% of all microorganisms are actually disease agents. The rest are quite necessary to the water pollution control process. Many are highly beneficial in that they cycle resources and energy through the natural environment. What is required of the worker is knowledge of and respect for microorganisms.

REFERENCES

1. Fradkin, L. et al. 1989. Municipal wastewater sludge: the potential public health impacts of common pathogens, *J. Environ. Health.* 51(3):148–152.
2. Gerardi, M.H., A.P. Maczuga, and M.C. Zimmerman. 1988. An operator's guide to wastewater viruses, *Public Works.* April, pp. 50–51.
3. Holvey, D.N., and J.H. Talbott, Eds. 1972. *The Merck Manual of Diagnosis and Therapy*, 12th ed. Merck, Rahway, NJ.
4. Vaughn, J.M., E.F. Landry, and M.Z. Thomas. 1983. Entrainment of viruses from septic tank leach fields through a shallow, sandy soil aquifer. *App. Environ. Microbiol.* 45(5):1474–1480.
5. Clark, C.S. 1987. Potential and actual biological related health risks of wastewater industry employment. *J. WPCE.* 59(12):999–1008.
6. Colwell, R.R., Ed. 1982. *Microbial Hazards of Diving in Polluted Waters*, Maryland Sea Grant Publication No. UM-SG-TS-82-01. University of Maryland, College Park, MD.
7. Pelczar, M.J., R.D. Reid, and E.L.S. Chan. 1978. *Microbiology.* McGraw-Hill, New York.
8. Kudo. 1971. *Protozoology*, 5th ed. Charles C. Thomas, Springfield, IL.
9. Hammer. 1975. *Water and Wastewater Technology.* John Wiley & Sons, New York.

8

Human Reproduction and Chemical Exposure

Recent times have seen an increasing awareness of the role the environment may play in the origin of certain birth defects. Birth injuries and defects are a major health problem in the world today. The World Health Organization estimates that 10% of all the human population may have a birth defect of some kind. The March of Dimes estimated in 1967 that perhaps 5% of the world population may be affected.[1] The U.S. Department of Health, Education, and Welfare estimates the rate of birth defects to be about 8%.[2] About 1% of the population may be severely disabled by the more serious birth defects. With the increasing participation of women in the water quality and engineering fields, many employers are acutely aware of the possible dangers that may exist in the field and lab.

However, it is a common error to assume that chemical agents only affect the woman. Unjustly, many employers in other industries have sought to bar women from employment rather than to correct the danger in the workplace. In certain industries, discrimination has become the justification for what is poor environmental housekeeping in the first place. One outstanding example of this is the attempt of employers in lead-processing companies and battery companies to bar women from the workplace;[3,4] yet an excess of lead exposure can affect the man's reproductive health as well.[5] In 1975 a study revealed that 150 men who worked in lead factories showed abnormalities in sperm structure.[6] Not only that, men who do not practice good hygiene in the workplace may actually bring home lead dust to the family.[7] Although the data appear soft or spotty in some areas, there is some evidence that lead exposure to the male can cause him to sire abnormal children, or the pregnancy of the wife of the worker may not be carried to term.[6] Poor hygiene endangers not only the worker, but also the family. Recently (March, 1991), the U.S. Supreme Court made the decision that one cannot be barred from the workplace, based on reproductive status alone. It is hoped that this court ruling will enable employers to address the real issue of chemical hazards, that is, how to make the workplace safe for all workers.

This is not to minimize the hazards women face during pregnancy. Certainly, there are special precautions that must be taken during pregnancy. However, the author does not believe that this always entails losing a job. By careful adherence to hygienic rules in the workplace, good nutrition and rest, and regular checkups by the prenatal doctor, a woman can work safely in many kinds of jobs during pregnancy. The key is safe work practices and attention to cleanliness in the workplace. More will be said about the special precautions the water quality mother should take.

8.1 SOME IMPORTANT TERMS

In order to understand how chemicals may affect the reproductive health of both men and women, one needs to understand some terms. One also needs to understand that mechanisms of induction can vary for the type of defect seen. Chemicals vary widely both in their potency and mode of injury to tissues and genetic material. With these points in mind, some basic mechanisms of reproductive injury will be covered.

There are two broad classes of reproductive hazards. The mutagens or mutagenetic agents actually destroy the DNA of the sperm or ova. Thus, a mutagenetic birth defect not only affects the child at birth, but that child can also pass the characteristic to HIS children. Examples of mutagenetic agents are mercury, which causes an interference with the mechanism of miosis; ionizing and microwave radiation, which cause breaks within the DNA helix itself; and nitrous oxide, which causes specific point mutations in human sperm cells.

Teratogenetic agents only injure the somatic tissue of the developing fetus. Thus, these defects are not passed down to the next generation. Birth injuries caused by this route vary widely in their severity. They can be extremely mild, i.e., low birth weight with no other symptoms, or quite severe, such as profound mental retardation. The word teratogenetic is derived from the French root Terat, meaning "monster." Agents that are teratogenetic include lead and mercury, which can accumulate in the fatty tissue of an infant's nervous system and placenta, and carbon monoxide, which causes placental respiratory insufficiency.[6,5] There is a point that should be made here concerning reproductive hazards: scratch a mutagen and you will find a carcinogen underneath.

Why is this so? There are two kinds of cells that can grow wildly, migrate extensively, and change their nature profoundly, in the course of development. These two kinds of cells are fetal cells and cancer cells.[8] Eventually the fetal cells specialize and settle down to a more reasonable rate of growth. Cancer cells, on the other hand, do not settle down and do not specialize. They continue to grow wildly. One theory of cancer induction is that certain genes that are only meant to be activated while a fetus grows and develops are activated again in the adult cell by certain chemical agents. Thus, good hygiene in the workplace may not only prevent birth defects but cancer as well.

When genetic material is damaged, there are a number of structural changes that can occur. In a point mutation, damage occurs to only one gene, which is in a group of three nucleotides. Chromosomal breaks are the actual severing of a chromosome or a group of many genes.[9] The break may involve only one side of the double helix (a chromatid break), or it may involve both arms of the chromosome (isochromatid break). The double break is the most serious, since a remaining template or blueprint of DNA is not left for repair. Human beings have some limited ability to repair certain damage in the genetic material if it is not too extensive.

8.2 FACTORS OF EXPOSURE

There are five factors affecting the severity of genetic damage to a cell and thus to the person:[5]

1. Age and health — Younger workers who have their family life ahead of them have a greater likelihood of being exposed over a number of years to injurious agents. For instance, men exposed for several years to nitrous oxide in careless atomic adsorption work could show severe damage to their sperm, resulting in sterility or the siring of a birth-defective child. Related to age factors is, for the woman, the stage of pregnancy. The stage of pregnancy controls what type of, and the severity of, the defect. The first trimester of pregnancy is the most critical and the most sensitive stage to chemical agents. Most of the development of a child occurs in the first three months of pregnancy. Both men and women in poor health or having bad personal habits have a greater chance of aggravating work exposure-related injuries. Smoking is synergistic with any pulmonary irritant such as carbon monoxide or nitrogen oxides.
2. Nutrition — A poor diet is damaging for men or women, but diet is most important for a woman who is carrying a child. It is known that nitrous oxide destroys vitamin B_{12}. It is theorized

that the teratogenetic damage to the child from such a gas is a result of inhibiting the use of this vitamin.[5] Thus, keeping the diet rich in B-complex vitamins is important. Even when a woman is not working during pregnancy, she should be encouraged to obtain good prenatal care. Directions given by the doctor on nutrition should be followed to the letter.
3. Dosage — It goes without saying that for both male defects in the sperm and female reproductive damage, the larger the dose of the offending agent, the worse the effect. All water quality workplaces are required to follow recommendations set out by the MSDS sheets for chemical reagents. All workplaces should have a safety program in place. So it is hoped that dosages of any offending substance will be kept below the 8-h TWA value. Certain chemicals and agents have NO safe threshold dose, however. Radiation is a good example of this. So is beryllium. Agents known to have no or extremely low threshold limits should be handled with the greatest care by both sexes.
4. Duration and number of doses — This is an area that is hard to predict, but there is some evidence that a single high dose may be more harmful to a woman than small chronic doses.[5] This is because the body is shocked by a large dose, but may have time to mobilize defense and detoxification mechanisms against chronic low-level exposure to an agent. Men, on the other hand, may experience more sperm damage with continuous chronic exposure. For instance, male workers exposed to Kepone and 1,2 dibromo 3 chloropropane, while keeping grounds free of fire ants, show testicular damage.[5]
5. Route of administration — As stated before, toxic agents can get entry to the body by ingestion, skin exposure, or inhalation. Coupled with this is the role the placenta plays in pregnancy. This organ, which originates from the tissue of the child, may either block the passage of certain chemicals or enhance the absorbance of other chemicals. It all depends on the molecular structure of the agent in relation to the biochemical structure of the placental wall. For instance, both mercury and most organic pesticides are actually preferentially absorbed by the placenta. Lead has greater difficulty in passing through the placental barrier, although enough gets through to cause damage. Most microbial agents are blocked by the placenta. However, there are a few microorganisms known to cause real trouble in pregnancy, and more will be said about these latter.

8.3 HAZARDS IN THE LAB

The mutagenetic and teratogenetic agents will be classed and described by chemistry. Effects will be covered for both men and women. Lastly, suggestions for a safer workplace will also be given.

Heavy Metals

There are several heavy metals that can induce birth defects. However, four metals are specifically known for causing not only birth defects, but other problems as well. These metals are lead, mercury, chromium, and nickel. Of these, nickel and chromium are more of a threat as carcinogens than actual agents of reproductive damage. They are included in this discussion because it is believed that certain cancers result from DNA damage.

Nickel's TLV is set at 0.1 mg/m^3 for an 8-h workday.[10] Most ventilation systems used in atomic adsorption should make meeting this requirement easy. Excess exposure to nickel, as with chromium, can induce lung and sinus cancer and also cause a contact dermatitis.[10,11] Nickel is used for making certain alloys and cheap jewelry, as is chromium. Thus, it is quite possible that it will be found in polluted water samples. Both men and women can be affected equally with these kinds of cancers. Although nickel has been found to cause reproductive damage in animals, it has not been proven to be a direct hazard to humans.[6] It is not known with what ease nickel can pass through the placental barrier. To date, no birth defects have been directly attributed to nickel, but it is recommended that the worker exercises all the usual cautions when running atomic adsorption analysis. Women carrying children should wear gloves, run the hood at its highest setting, and use a lab coat during the preparation and analysis of a water sample. Lab coats should be left at work and laundered separately from all other household clothing. It would not be a bad idea for men to keep all work clothing completely separate from the family laundry as well. There have been well documented cases of men carrying a chemical poison home to the family on his work clothes.

Chromium's action and hazards are much the same as nickel. The same precautions used in handling nickel should therefore be used for chromium. This metal IS needed as a nutrient, but the dosage is small, less than 100 µg/d. NO pregnant woman should ever take chromium, or any other vitamin supplement, without the express knowledge and direction of her doctor.

Lead and mercury have been documented extensively in the literature for several debilitating effects. Lead has a TLV of less than 0.15 mg/m^3. It is not only a problem in industry, but it is found practically everywhere. Old houses painted with old-type paints have the potential of poisoning anyone trying to restore them. Many do-it-yourself books warn that the home owner should wear a mask when restoring old surfaces and should wash thoroughly after the job is done. It is estimated that 1 to 5% of all house dust may contain lead.[9] However, with the use of nonleaded gasoline, this figure may be falling. Some epidemiologists have even stated that, in the past, enough lead was deposited into city atmospheres to cause subclinical lead poisoning in many children living in those areas. One, therefore, does not have to work in a battery company or a foundry to be exposed to lead. With public awareness of the problem, and improvements in paint and gasoline purity, the ambient lead problem is slowly being brought under control. In atomic adsorption work there is simply no excuse for lead poisoning. With proper precautions it should be easy to maintain an atmosphere of less than 0.15 mg/m^3 of lead.

The effects of lead can be felt throughout the life of the exposed person, but it is the effect lead has on human reproduction that causes the most concern. In the man, lead damages the sperm. Sperm exhibits poor motility, low density, and actual chromosomal damage.[6] As early as 1860 a study showed that wives of lead workers lost more children within the first year, had more miscarriages, and had a greater difficulty in getting pregnant than did the general population.[6] Women themselves can be the primary target of lead toxicity. In women lead can cause disruption of the menstrual cycle. In pregnancy lead can hamper the implantation of the blastocyst on the uterus wall. Even small amounts of lead can damage the central nervous system of the developing fetus. If lead is kept below the level of toxicity for the woman, there is strong evidence that the fetus will be protected.[6] A clean, well-dusted lab goes a long way in avoiding problems with lead exposure for both men and women.

It is quite possible that mercury will replace lead as a common hazard both at work and home. Mercury is replacing lead as a paint preservative in certain house paints. The home vapor danger from painting is very real. Now some manufacturers are replacing their mercury preservatives with a safer substitute. In industry, mercury is found in the production of electrical appliances and vapor lamps. In water quality work the major exposure to mercury is in cold-vapor atomic adsorption analysis. The TLV of mercury is set at 0.05 mg/m^3 for metallic vapor and 0.001 ppm for alkyl compounds.[10] The effects of mercury on both males and females is dramatic. Mercury directly attacks genetic material. Its specific mode of action is to damage the spindle formation of the cell during miosis. Thus, the genetic material cannot be properly distributed between the two daughter cells. In men this results in sperm damage. In women this can damage the ovaries. Both sexes can suffer CNS damage. Mercury also is preferentially adsorbed by the placenta and the fetus. Mercury tends to accumulate in fatty tissue, the gonads, the pituitary, and the hypothalamus of the brain.[6] Thus, the developing fetus can suffer severe nervous system and hormonal damage. Mercury has, as stated before, a biological half-life of about 70 d. It is not easily excreted by the body. Thus, it behooves both men and women to insist that proper air filtration and ventilation be used when running cold-vapor analysis, COD's, and TKNs. NO mercury compound should be handled without a gas hood that can run at least 75 to 100 fpm at the entrance of the hood. Mercury detectors should be placed at strategic points in the preparation and digestion area. Always wear gloves when handling mercury and its related compounds. Any clothing used on the day of mercury work should be washed completely separate from the family wash. With a strict adherence to lab safety and hygiene rules, mercury problems can be controlled.

Gases

Any gas, through its asphyxiating action, can damage a fetus. Just one incident with carbon monoxide can cause severe mental retardation.[12] It is important that the flames used in atomic adsorption be well ventilated. All digestion procedures should be run under a hood. Acids should be handled in only well-ventilated areas. Periodically the lab should be checked by a professional industrial hygienist for dead zones of air. Corrective measures should then be undertaken.

It is not likely that carbon monoxide, chlorine, hydrogen sulfide, or other reactive gases will be a problem in the chemistry lab, since most labs have well-equipped gas hoods. There is one gas, however, that gives concern. Nitrous oxide is used in atomic adsorption analysis for such refractory elements as chromium, aluminum, and titanium. The higher temperature of the nitrous oxide oxidant is needed to break up a chemical sample into its basic atoms for analysis. The effects of this gas on both men and women have been well documented in the medical fields. Anesthesiologists have reported increased birth defects and miscarriages, as well as sperm damage. This has not been well documented by technicians using this gas in atomic adsorption, however. The problem is probably not as serious, since the nitrous oxide is actually being consumed and destroyed in a flame. Yet the chemist should be aware that faulty gas hoods, poorly connected nebulizer parts and/or gas lines, and faulty regulators can be a source of leaks. Atomic adsorption equipment should be checked frequently and repaired as needed. The National Institute for Occupational Safety and Health recommended in 1977 that nitrous oxide levels never reach levels above 25 ppm.[13]

In men some studies show that nitrous oxide can cause reversible damage to the sperm.[6] Some authorities differ about the degree of severity of damage to men by nitrous oxide.[5,6] But more severe damage can occur in women. This is one agent that seems to affect women more adversely than men. Female dental assistants show more infertility than does the general population. The miscarriage rate is also higher in this group of women compared to the general population. The number of deformed children born to dental assistants and nurses exposed to nitrous oxide is also higher than in ordinary people. It is important to realize that a lot of the dental offices and operating rooms had nitrous oxide levels higher than 130 ppm, and in some cases as high as 700 ppm.[13] From this data one sees how important proper ventilation and maintenance of equipment are. As long as supervisors are careful to see that the work area around the atomic adsorption equipment has less than 25 ppm of nitrous oxide, there should be no damage to either men or women working in the area. Good nutrition is also important for the worker. It is necessary to have adequate amounts of vitamin B_{12} in the diet, especially when working with nitrous oxide.

In the field the three gases of concern are hydrogen sulfide, carbon monoxide, and ozone. Carbon monoxide and hydrogen sulfide affect the fetus primarily by obstructing the flow of oxygen across the placenta. As is documented for pregnant smokers, carbon monoxide can cause retardation of growth and, thus, low birth weight.[2,5,6] Hydrogen sulfide is suspected of causing a higher rate of spontaneous miscarriages. Thus, ventilation is extremely important around wet wells, garages, and pump stations. It seems that most evidence points to women being more affected by these gases than are men. But there is some evidence that hydrogen sulfide might cause damage to the sperm.[6]

Ozone, which is becoming more popular as a disinfectant, has been suspected of causing DNA damage in both men and women. Some flight attendants in the airplane industry claim that they have had higher rates of birth defects and miscarriages than casual flyers. This has yet to be documented by independent studies.[5] A study has shown that DNA damage occurs to human cell cultures exposed to 8-ppm ozone.[5] This is far above the 8-h TLV of 0.1 ppm.[10]

Organics

Almost all insecticides and herbicides have been documented as having some mutagenetic or teratogenetic effects. All of these agents tend to be fat soluble. Thus, once in the body, they tend to stay there for a long time. For instance, organochlorine insecticides may have biological half-lives ranging from a few months to years. DDT has a field half-life of about 20 years.[9] It has been found even in the tissue of animals in remote wildlife areas.[9] Agents like the herbicide 2,5,5 trichlorophenoxy-acetic acid have been suspected of causing damage to both the female and male reproductive systems.[5] Effects of pesticides can range from premature birth, to birth defects involving the CNS and internal organs. Some agents, such as agent orange, are suspected of

There are too many priority pollutants to try to describe what action each has on human reproduction. Indeed, many of the organic pollutant actions on human reproduction are not well understood. Because not much is known about the organic chemicals, it is wise that both sexes exercise the utmost caution in handling such agents. As stated before, all gas chromatographic standards should be prepared under an enclosed glove box, as these are usually extremely pure and concentrated extracts. Protective gloves and clothing should be worn while working with such standards.

When working with pesticides on the grounds, every precaution should be taken. The author recommends that, if possible, any woman carrying a child avoid working with any kind a pesticide while maintaining the grounds of the plant.

Radiation

There are few sources of radiation hazards in water and wastewater treatment, but the water quality employee should be aware of some of the possible hazards connected with radiation exposure. There are two broad classes of radiation that can affect lab workers. Ionizing radiation includes gamma rays and high-energy particles such as electrons, protons, and neutrons. The longer-wave radiation includes microwave radiation, and disturbances from high-power electrical fields.

It has been suggested that, for both women and men, there are NO threshold levels of safe exposure when dealing with radiation, that is, even very low levels may have some effect on the body.[14] It is even suggested that the stray cosmic rays that reach the earth are responsible for some of the random mutations in both human and animal populations. Radiation in the water quality workplace can come from electron capture detectors that are used in gas chromatography. They can come from the microwave digesters used in the wet ashing of certain samples. In field hydrology there is the possibility of having to use a neutron or gamma radiation source, such as cobalt 60 or a lead-beryllium material, in logging geological formations while drilling a well.[15] Very rarely a wastewater analyst may be asked to analyze a sample for radioactive contamination. Those people who monitor wastes from nuclear power plants and isotope processing plants, need to make sure that proper procedures and shielding measures are followed. Nuclear chemistry is a specialized field in itself. If a water quality lab is required to handle radioactive samples, it is strongly suggested that these samples are sent out to specialized labs or that an expert consultant be hired in the setting up of a lab for radioactive samples.

The most common source of "shortwave-frequency" radiation in the lab is a microwave digester. These units are nothing more than special microwave ovens used to digest certain metal and nutrient samples.

Microwave exposure has a number of effects. The sperm can be damaged in men by microwave exposure. Some people claim that men who work with radar or other microwave sources have a higher incidence of birth defective children. However, the U.S. Navy could find no definite correlation between exposure and adverse effects on the male reproductive system.[6] In women, microwaves can affect the CNS of the developing fetus. It is recommended that leakage from microwave digesters be kept below 1 mW/cm^2.[6] Most microwave ovens and digesters are so constructed to have leakage below this level. One can have their digester checked by a local health department or can obtain a fairly inexpensive microwave meter for safety testing. Certainly, one should exercise the same precautions with a laboratory unit as they would with the home microwave oven. Have units checked once a year. Do not stand directly in front of a working unit. Leave the room during the digestion process. Keep the digester clean and in good repair. Make sure the seal is tight when the door to the unit is closed. Replace faulty seals. Ovens that are in poor condition or are outdated should be replaced.

Exposure to particle or ionizing radiation has well-documented effects on both male and female reproductive health. The degree of damage depends on the dosage, the length of exposure, and the type of radiation. Neutron sources are the most destructive because they further decay into protons and electrons, with some gamma radiation, once they are lodged deep into the body. Beta (or electrons) rays and alpha rays (a naked helium nucleus) are the least damaging. These particles either have a small area of influence or low penetration into the body. All of these particles can cause chromosomal breaks. Men's sperm can be damaged with as little as 15 rad (adsorbed dose of 100 ergs/g: think of 1 erg as the energy needed to push 1 g about 1/1000 of a cm).[6] Women, during pregnancy, can suffer severe damage to the developing fetus. Multiple defects of the endocrine system and the CNS can result from radiation exposure. The degree of damage is most extreme during the first trimester of pregnancy. As little as 1 rad can affect the developing fetus. For purposes of safety, rads are converted to a unit called a rem (rem = rad × QF × DF). QF is a rating of a nuclear particle. For instance, electrons are rated 1, while neutrons have a rating of 3 to 10. DF is the dose factor that has to do with how the radiation is distributed within the body. For example a stream of electrons could give a rating of 1 rad × 1 × 1, to give 1 rem. A stream of neutrons might be 1 rad × 3 × 1, to give a value of 3 rems. NO worker should ever be exposed to more than 5 rems a year, and no more than 3 rems should be allowed within a 13-week period. Consider that in the course of a year a person may be exposed to no more than 55 millirems a year in the course of getting tooth and chest Xrays.[10]

That is the bad news. The good news is that most pollution personnel probably will not be exposed to even 1 rem/year. Most of the particle radiation that a water quality chemist may be exposed to comes from electron capture detectors used in gas chromatography. The radiation of an electron capture detector is an extremely weak source of beta or electron radiation. A piece of ^{63}Ni foil is the source of the electrons. It is located deep within the structure of the detector chamber. Less than 10 millicuries (mCi) of this isotope is used.[16] A curie is 3.7×10^{10} atomic disintegrations a second. These are low-energy electrons that probably cannot go much beyond the skin barrier. Still, common sense dictates the handling of any gas chromatography and mass spectra equipment. The rules are as follows:

1. Do not attempt to repair the detector yourself. Hire a technician trained in this work.
2. Wipe test the gas chromatograph at least once a year. The Atomic Energy Commission provides kits for testing for stray contamination, and the directions for their use.
3. Follow directions in setting up the gas chromatograph to the letter. Do not alter the suggestions of the manufacturer.

With good care the electron capture gas chromatograph should not pose a threat to the worker.

It would take a separate book to cover all aspects of radiation safety. Work with nuclear sources or pollutants demands special training. If there is going to be a great deal of work with high-radiation sources, it is strongly recommended that the technicians be encouraged to take a "health physics" course at a local college. If such a course is not locally available, it might be wise to bring a specialist in for training the workers.

Microbial Hazards

Although a fetus is fairly well protected from infection by the placenta, there are a few diseases that can be a cause for concern. A bad case of flu or a cold can cause spontaneous miscarriages, or if the fetus survives, defects and tumors may result.[2] Thus, some of the enteric viruses are a source of worry and concern. Chicken pox, mumps, hard measles, and rubella can cause deafness, blindness, heart defects, and mental retardation. The good news is that disease agents are teratogenetic in their effects. They cannot be passed from generation to generation. Another good piece of news is that mumps, hard measles, and rubella have vaccines that can be used for protection. It is suggested that women carefully check their medical records to see that all vaccinations are up to date. Any vaccinations should be given WELL before even thinking about becoming pregnant. Probably, a lead time of three to six months should be allowed between vaccination and conception. A woman should check with her doctor as to the proper interval between vaccination and conception. Protozoan infections and internal parasites can pull down the health of the mother and thus endanger the fetus. The woman who works in bacteriology should be careful to protect herself from infection.

Men are not completely off the hook. A mumps infection can render an adult sterile. Typhoid, tuberculosis, and a bad case of flu can also damage sperm.[6] Men should have regular physical checkups and vaccinations.

Keeping the Lab Birth-Defect Proof

By now it must be apparent that the avoidance of reproductive hazards concerns both men and women. No one group or sex is solely responsible for the developing child. Chemical agents do affect men and women differently. But it should be evident that men are in as much danger as women in having their reproductive and family health impaired. Discriminatory action is not the answer. Good housekeeping practices are more appropriate than banning one or the other sex from the workplace. By instituting a good industrial hygiene program, the employer not only prevents birth defects, but also cuts down on lost time and medical costs caused by toxic exposure.

These actions are suggested for the reproductively healthy workplace:

1. Ask that all workers in the lab and field keep up with their vaccinations. Ask that all workers get a complete physical at least every two years. Most water pollution control centers provide medical insurance coverage, so it should not be difficult to institute a good preventive health program.
2. Encourage workers to give up smoking and drinking. The abuse of alcohol, tobacco, and other drugs have profound effects on the reproductive health of men and women.
3. By now it should be apparent that ventilation covers a multitude of sins. Inspect all gas hoods, fans, and blowers. Make sure that all ventilation equipment meets suggested safety standards.
4. Where one is dealing with especially toxic substances, one may want to take special precautions. For instance, both a carbon trap and a gas hood should be used when testing for mercury.
5. One cannot avoid the fact that a pregnant woman is especially vulnerable in the workplace. There are certain activities it would be wise for her to avoid. Work with mercury and pesticides

should be avoided. Bacteriology work is safe if the woman wears gloves, a mask, and a protective lab coat. Strict aseptic procedures should be followed at all times.

6. A woman that is pregnant should inform her gynecologist of the type of work she does. The doctor should also be aware of the agents that the woman may have to handle. Regular prenatal care is important for any woman. The last word in what a woman may or may not do in pregnancy should come from her gynecologist/obstetrician.
7. Both men and women should have uniforms or lab ware. This special clothing should only be worn at work. They should be washed separately from all other family laundry. It has been well documented that agents such as lead can be brought home on work clothes.[7] If the job a worker must do that day is particularly odious, he or she may also want to shower at the plant before coming home. Many plants do have shower facilities.

With forethought and cleanliness, the risk of exposure to mutagenetic or teratogenetic agents can be greatly reduced. Proper protection in handling chemicals is important. Safe procedures should be followed at all times, even when doing such simple lab analysis as pH or BODs.

REFERENCES

1. Kennedy, W.P. 1967. Birth Defects Series: Epidemiologic Aspects of the Problem of Congenital Malformations, vol. 3. no. 2. National Foundation of March of Dimes.
2. Wilson, J.S. 1981. The Knowledge and Behavior of Pregnant Women in Relation to Teratogenetic Agents and Teratogenetic Behaviors. Master's Thesis. Kent State University, Kent, OH.
3. Goodmen, E. 1991. Quest for equality in the workplace (editorial), Akron Beacon Journal, March 28, Akron, OH.
4. 1991. Ups and Downs of Discrimination. *U.S. News.* April 1. p. 10. .
5. Kurzel, R.B., and C.L. Cetrulo. 1981. The effect of environmental pollutants on human reproduction, including birth defects. *Environ. Sci. Technol.* 15(6):626–640.
6. Gibbons, J.H. 1986. *Reproductive Health Hazards in the Workplace.* 99th Congress. U.S. Office of Technology Assessment. January.
7. Lippmann, M. 1990b. Lead and human health: background and recent findings, *Environ. Res.* 51:1–24.
8. Beaconsfield, P., Birdwood, and R. Beaconsfield. 1982. The placenta, *Sci. Am.* 243:94–101.

9. Waldbott, G.L. 1978. *Health Effects of Environmental Pollutants*. C.V. Mosby, St. Louis.
10. Proctor, N.H., and J.P. Hughes. 1978. *Chemical Hazards of the Workplace*. J.B. Lippencott, Philadelphia.
11. Stich, H.F., Ed. 1985. *CRC Carcinogens an Mutagens in the Environment: The Workplace*. CRC Press, Boca Raton, FL.
12. De Vinck, C. 1985. The power of my powerless brother, *Reader's Dig*. July.
13. Schumann, D. 1990. Nitrous oxide anaesthesia: risks to health personnel. *Int. Nursing Rev.* 37(1):214–217.
14. Cember, H. 1976. *Introduction to Health Physics*. Pergamon Press, New York.
15. Fetter, C.W. 1980. *Applied Hydrogeology*. C.E. Merrill Publishing, Columbus, OH. p. 437.
16. H.H. Willard, L.L. Merritt, and J.A. Dean. 1974. *Instrumental Methods of Analysis*. D. Van Nostrand, New York. p. 530.

9
Setting up a Safety Program

With awareness of existing dangers in the workplace, it should be relatively easy to correct the problems. Setting up a sound safety program requires a little research and forethought.

One of the things a water or wastewater plant should have is a good basic safety library. All MSDS sheets for chemicals, paints, cleaning agents, and pesticides should be kept in a hard-bound notebook. The notebooks should be kept in the central study area in a prominent place of easy access, such as the security department or the lunch and break area. In our case the lab is the most open area because of the need for testing sludges on all three shifts. So the safety material is kept there. There should also be at least one good basic safety book. The Red Cross can supply a good basic first aid book. All workplaces should include in their library these references:

1. The CRC *Handbook of Laboratory Safety* — This is a classic. It covers just about any situation a worker may encounter in the lab.
2. *Chemical Hazards of the Workplace* by Proctor and Hughes — This book is geared more for the use of an industrial doctor or nurse. It is quite advanced, but supplies much information that is highly specific for many kinds of chemicals. Also, this book has an excellent section on the detection and correction of dangers.
3. A good book on physiology and toxicology — There are several good sources. It is up to the employer to select the reference most suited to the needs of his workers.
4. If the lab is to work with radioactive hazards — The author strongly suggests getting a good basic health physics book.

5. The Nutrition Almanac by J. D. Kirschmann — By now it should be evident that nutrition can affect the way a worker responds to chemical injury. It would not hurt the worker to review his basic health and hygiene from high school days.
6. The Merck Index of Chemicals — This has good, concise, decriptions of most industrial chemicals.

A good basic library encourages the worker to review important concepts of health and hygiene.

Simply reading about safety is not enough. Training is important. A basic series of training courses should include first aid, CPR, and training with a self-contained breathing apparatus. Where indicated, specialized training courses may be needed: for instance, in the use of specialty gases in atomic adsorption work or gas chromatography. Other areas that need specialized training include any work involving pesticides or radioactive materials.

Courses need not always be expensive. The local Red Cross should be able to help with some of the basic safety training. Water and wastewater courses are now also including sections on safety.

Remember that training is not a one-time event. It is important to refresh a worker's safety knowledge from time to time. Also, retraining keeps the worker sharp on the job. Every person tends to become careless as time goes by. Refresher courses remind a person to take care on the job.

9.1 BEING PREPARED

Even with the best of programs, accidents can happen. Effectiveness in dealing with emergencies is directly related to the degree of preparation in the safety program. Very large plants or industrial pretreatment facilities within a large industry may have their own doctor and nurse on the property. Most sewage treatment plants and water treatment plants are too small to be able to afford full-time industrial hygienists. Therefore, it is important for at least two people on each shift to be well trained in safety and first aid. It is also important for no worker to be completely isolated from fellow workers when doing a task. In some work the buddy system is very important. When working with cyanide determinations or pesticides, for example, two people should always be in contact with each other. When working overtime the worker should clock in and out with a shift supervisor. In this way it is always known just who is working at a dangerous task at all times. A very good illustration of this involves the time a worker was rescued from asphyxiation simply because he was late for a meeting. It was known exactly where he was scheduled to work and at what time. The superintendent was able to check the exact manhole immediately, and the man was rescued. In this case, detailed scheduling and the awareness of at least one other person regarding the whereabouts of a worker saved a life.

In addition to keeping current with ventilation requirements of the workplace and having well-equipped fire stations, good first aid stations are needed. A good first aid station should be located close to the drench shower or eye shower. It should contain these items (see Figures 9.1 and 9.2):

1. One burn kit — Enough to cover extensive areas of the body
2. A heavy-duty large industrial first aid kit
3. Suitable neutralizing solutions for acidic and basic burns

The kits should be inspected and updated frequently. In the author's plant, a service has been hired to maintain and update the main first aid station. Smaller first aid kits should be kept in the lunchroom, offices, and other prominent locations around the plant. Make sure every self-contained breathing apparatus is full of clean air and in good repair. Only purchase compressed air for such kits at a reputable SCUBA diving shop. DO NOT fill the apparatus with compressed air from an industrial compressor. Serious lung injury and lipiodal pneumonia can result from breathing dirty compressed air. If the workplace wishes to purchase their own compressor or use one from the local fire department, it must meet diving air supply specifications.

A well-stocked first aid station allows the rescuer to supply the proper first action for an injury. Effective treatment in the first minutes of an injury can greatly reduce the severity of damage to the victim.

9.2 WHERE TO LOOK FOR NEEDED INFORMATION AND CARE

There is only so much a person can do in the first line of emergency care or spill cleanup. Some injuries will require extended care by a specialist. Some spills or old chemicals are far too dangerous to remove without the aid of a special waste group. Therefore, it is important for the superintendent and workers to know who to call and where to go in case of an emergency.

As was stated before, the water quality workplace should work closely with the local fire department. The local department should be aware of the exact nature of the work done and the chemicals used. Know where the bomb squad is in your area. Our Summit County is fortunate. Akron's fire department also runs the bomb squad. Just how important this is was demonstrated when the workers were assigned

110 CHEMICAL HAZARDS AT WATER TREATMENT PLANTS

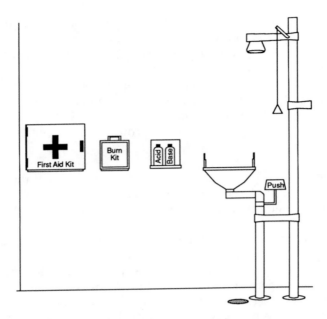

Figure 9.1 A safety and first aid station. A good station should have an eye bath, a drench shower, a burn kit, and a first aid kit. Neutralizing solutions should be kept fresh. Renew them at least once a week.

the cleaning and sorting of a chemical storage area. Some old ethyl ether was found in a tin. A chemist promptly stopped work, read the *Merck Index*, and informed the superintendent, who promptly called the Akron bomb squad. As a result an outdated and possibly explosive chemical was removed and neutralized safely. We in this location are fortunate for another reason. The Akron Children's Hospital is the location for both the local burn and poison centers. A list of important numbers that should be posted in a prominent view area include

1. the local fire department
2. the sherrif's office or bomb squad
3. the medical center that specializes in burns and poisons
4. the closest hospital location for routine medical care
5. a doctor who specializes in environmental medicine

Look carefully in the local newspapers during Earth week or at other environmentally important times. The *Akron Beacon Journal* in Akron, OH did a marvelous job in giving a list of important numbers for Ohioans to contact in the event of an emergency. The local fire depart-

SETTING UP A SAFETY PROGRAM 111

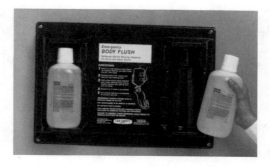

A

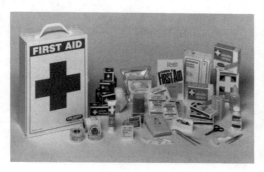

B

C

Figure 9.2 Some well-stocked first aid kits. Make sure that the lab safety station has the means to deal with chemical injuries, burns, insect bites, poison ivy, cuts, and mild sickness such as colds or flu. (Photos courtesy of Lab Safety Supply, Inc., Janesville, WI.)

ment may also be able to provide additonal information on who to call for certain emergencies. It is important for each workplace in every state to develop an effective network of support services to be used during an emergency. Here is an example of just such a list that applies to Ohio:

1. Attorney General's Office — To report illegal dumping (800)-282-3784
2. Main Ohio EPA Office in Columbus — (614)-644-3020
3. State Fire Marshall — (800)-282-1927
4. Ohio Cancer Information Service - (800)-422-6237
5. Public Utilities Commission of Ohio — This is a number for reporting major industrial spills (800)-686-8277
6. Resource Center — Industrial Commission of Ohio Division of Safety and Hygiene (800)-282-3045
7. Summit County Disaster Services — Mr. Scott Searingen, Coordinator, at (216)-379-2558

Every technician, operator, and superintendent is responsible for becoming informed about what to do and who to call during an emergency.

9.3   THE PLAN

At various points in this work, exact procedures for caring for a certain injury were given. Because the first response to an injury is so important, some general first aid rules will be reviewed. When a person is injured, the first three courses of action that should be taken are remove, resuscitate, and neutralize. Certainly, before any other action, the person must be removed from the danger. But care must be taken in protecting oneself. For instance, one should never enter a chlorine leak area without a proper breathing rig. Quickly get the person clear of the danger before beginning first aid. If it appears that resuscitation is needed, shout to someone to call the ambulance. Begin CPR or artificial respiration immediately if needed. Also, treat the victim for shock if indicated. Stop any bleeding with direct pressure on the injury. Next, when the person's breathing is stabilized or the bleeding has stopped, one must turn attention to neutralizing any chemical injury present. This may only require generous application of water to the injured site, or in the case of basic or acidic injuries, one may need to apply a neutralizing agent after the first rinsing with water. When helping the injured party, make sure to sound the alarm to another worker, so that the ambulance can be on the way. The less time one can put between the injury and a visit to the hospital, the better the chance for a person's complete recovery.

Some injuries are minor and do not require hospitalization. Small

SETTING UP A SAFETY PROGRAM 113

```
                        AN ACCIDENT FORM
                              ACCIDENT REPORT
        Name of employee_____Soc. Sec. No._____
        Home Address and Phone_____

        Employed by_____Department_____
        Job title of employee_____Dept. Phone_____
        Supervisor_____

        Location of the accident_____
        Nature of the accident_____

        Location of accident_____
        Time of the accident_____
        Did the accident occur during work hours_____
        To whom was the accident reported_____
        Name of witness_____Phone_____
        Home address of witness_____

        Name and address of doctor and/or hospital rendering service
        _____
        _____Phone_____
        Physician's report of accident and medical actions taken____
        _____
        _____

        Diagnosis and prognosis_____
        _____

        Additional comments_____
        _____
        _____

        I, the employee certify all the above statements are true
        Signature of employee _____Date_____

                              SUPERVISOR'S REPORT
        Nature of injury_____

        Remarks_____

        Did the employee report back to work?_____
        Date returned to work_____
        Date and time injury reported_____
        Was employee full or part time_____
        Supervisor's signature_____Date_____
        Office or department_____
```

Figure 9.3 A typical Accident form.

injuries can be taken care of at the first aid station. However, the worker should be encouraged to see his own doctor in order to make sure the injury does not progress to something more serious. ALL injuries, serious or not, should be reported on an "at work injury" form. This is important, as the extent of some accidents are not apparent at the time of occurrence. Worker's compensation cannot be paid without good documentation. Many forms of work injury documents can be used, (see Figure 9.3). The forms should include

1. the name, work position, and home address of the worker
2. time and place the injury occurred
3. a description of the exact nature and cause of the injury
4. a description of the extent of the injury
5. detailed documentation of the first aid actions taken and (if applicable) when the person was admitted to the hospital
6. a section reserved on the form for the doctor to fill out describing any follow-up treatments used

A well-run workplace that gives attention to hygiene and has an extensive safety program will save costs in the long run. Absenteeism due to injury can be dramatically cut. Workers also show a better attitude towards work if they have a clean and safe area for their tasks.

Appendix A
A Word about Toxicology

Because this is a book about chemical hazards, it would be useful to understand some of the basic concepts of toxicology. In order for effective first aid to be performed, it is necessary to understand something about the behavior of poisons in the body.

There are three basic ways a poison can enter the body: through the mouth, skin, or the lungs. Water quality personnel need to worry mainly about the skin and lungs. It is important to properly protect the skin and lungs when working with chemicals. The eyes, especially, can be injured even by fumes. Protective clothing and goggles must be worn in certain work situations. The route of entry can affect the severity and speed at which a toxicant acts. The worst route for many poisons is the oral route. Ingested poisons can act very rapidly. However, oral entry is usually not a problem in the water quality field. Lung entry can be rapid in its effects, but if quick action is taken, the dosage of poison absorbed by the lungs can be kept small. The main worry with lung damage is tissue injury and edema, which can result in respiratory problems later. The skin is an effective barrier against entry by many chemicals. However, certain organics and some metal solutions can easily cross the skin barrier. In the water quality field one should be acquainted with those chemicals used that can cross the skin barrier. Such skin-active substances should only be handled with gloves. Make sure policies for the safe handling of chemicals are set up and followed to the letter. Keep the workplace clean, in good repair, and properly ventilated.

Understanding the toxicity of a chemical requires some basic knowledge of physiology and dosage factors. A good understanding of units would be helpful. There are many conversion booklets on the market that can be quite helpful in doing calculations. Here are a few handy units that will aid in the understanding of this book:

- 1 mg (milligram) is 1 one millionth of a kilogram (kg), hence the term ppm, or parts per million.
- 1 kg (kilogram) contains 1000 g.
- 1 g (gram) contains 1000 mg.
- 1 mg contains 1000 µg (micrograms).
- 1 g contains 1,000,000 µg.
- 1 kg contains 1,000,000,000 ug, hence the term ppb, or parts per billion.
- 1 l (liter) has 1000 ml (milliters).
- 1 m^3 contains 1000 l.
- 1 m^3 contains 1,000,000 ml.

With these basic units under the belt, one can understand the notations used in the MSDS sheets.

There are two basic units of measurement used in the determination of exposure to a poison. One is the ppm or ppb unit. This is a weight-to-weight factor in water mixes and a volume-to-volume factor in gaseous mixes. For instance 5 ppm of sulfur dioxide means that for every million volumes of air, there are 5 volumes of sulfur dioxide. Another method of notation is the mg/l, µg/l, mg/m^3, or µg/m^3 unit. This is a weight-per-volume notation, that is, 1 µg or sometimes 1 g of substance per liter or cubic meter of air. Since this must take into account the density of air in relation to the density of the substance, a mg/l does not equal a ppm in gaseous mixes. So actually 5 ppm of sulfur dioxide is 5 volumes of the poison per 1 million volumes of air. In the mg/m^3. notation, that translates to about 13.07 mg/m^3 of sulfur dioxide to air at sea level and 25°C. However, in water 5 ppm translates very closely to 5 mg/l because the density of the poison/water mix is so very close to 1000 g/l (see calculations).

For gaseous mixes:
(This is at 1 atmosphere (atm) and 25°C.)

$$\frac{5 \text{ ml SO}_2}{1{,}000{,}000 \text{ ml air}} \quad \text{or} \quad \frac{5 \text{ ml SO}_2}{1000 \text{ } \ell \text{ air}} = 5 \text{ ppm}$$

So:

$$\frac{5 \text{ parts}}{1{,}000{,}000 \text{ parts}} \times \frac{64{,}060 \text{ mg}}{\text{mole SO}_2} \times \frac{1 \text{ mole air}}{24.5 \text{ } \ell \text{ air}} \times \frac{1000 \text{ } \ell}{\text{m}^3}$$

$$= 13.07 \frac{\text{mg}}{\text{m}^3 \text{ air}} \quad \text{or} \quad .013 \frac{\text{mg}}{\ell} \text{ air}$$

For aqueous mixes:

APPENDIX A—A WORD ABOUT TOXICOLOGY

(This is at 1 atm and 25°C.)

$$\frac{5 \text{ mg } SO_2}{1,000,000 \text{ mg } H_2O} \quad \text{and} \quad 1 \ \ell \ H_2O \text{ weighs } 997,045 \text{ mg}$$

$$\frac{5 \text{ mg } SO_2}{10^6 \text{ mg } H_2O} \times \frac{997,045 \text{ mg } H_2O}{\ell} = 4.98 \text{ mg}/\ell$$

This is roughly 5 mg / ℓ.

For very concentrated water mixes, the density of the solution must be taken into account. For example for a 30% solution of ferric chloride ($FeCl_3$)

$$\frac{300,000 \text{ mg } FeCl_3}{1,000,000 \text{ mg } H_2O} \times \frac{1,320,000 \text{ mg}}{1 \ \ell \text{ of solution}} = \frac{396,000 \text{ mg}}{\ell} \quad \text{or} \quad \frac{396 \text{ g}}{\ell}$$

If one looks at the literature, one can see that exposure levels are given in both kinds of notation.

There are some terms that have to do with not only the concentration of toxicant, but also the duration of exposure. These are the TLV-TWA and TLV-C notations. The term TLV means threshold limit value, the allowable concentration of a toxicant in a workplace. It can be combined with other notation to specify the circumstances of exposure. Note:

1. TLV-TWA means the threshold limit value-time weighted average for an 8-h workday or a 40-h workweek. This is the level considered safe for an average worker. Sometimes the TLV is dropped as understood.
2. TLV-STEL is the notation for short-term exposure limit. This is the concentration level that is only allowed for 15 min without injury. In other words, at this concentration the worker has 15 min to clear out before injury results.
3. TLV-C is that exposure that will result in injury even if exposed for an instant. The C means ceiling or critical.

There are several factors that affect the severity of a chemical injury. One concern is that workers are not just exposed to one chemical at a time. A mix of agents can act quite differently on the body than can a single substance. There are antagonistic and synergistic effects. Two or more agents can be said to be antagonistic if the effects of the mix are less severe than any one agent alone. For instance, ammonia gas and chlorine gas could be said to be antagonistic, since they combine to make ammonium chloride. This salt is less toxic than either of the gases alone. But a mix of poisons can be more toxic that either agent alone. Such a mix is said to be synergistic. For example, having alcohol in the blood stream can make a person more sensitive to exposure to organic solvents such as ether. A second aspect of synergism is cross-sensitization. Cross-sensitization can result when the body is affected by one chemical and then becomes hypersensitive to other related chemicals. For instance, exposure to benzene might sensitize a person to several other compounds that contain a benzyl ring. Such a person may show an allergic reaction to a related benzyl compound on the very first exposure.

In addition to these factors, the state of health can affect the severity of response to a toxic chemical. Persons can have different responses to the same concentration of a substance, due to other circumstances. Changes in baseline health result in changes in response to an injury. When treating an injury, one must consider the age of the injured party, his or her basic health, how that person was exposed (was it an oral, skin, or a lung route), the length of exposure, and the concentration of exposure.

When reading MSDS sheets one will notice a numerical code describing the degree of risk in using the chemical. The labeling for a particular chemical might go something like this: health 3, reactivity 1, flammability 3. For this chemical, this means that is it quite toxic, mildly reactive, and highly flammable. The higher the number, the worse the hazard. Before working with any substance, the worker should know its hazard profile. With a basic understanding of some toxicological terms, the worker can understand and use the MSDS sheets to his or her benefit.

Glossary

Acid — A substance that increases hydrogen ion concentration in a solution. A proton donor.

Aerobic — A process requiring oxygen. For example, activated sludge treatment is an aerobic process.

Addictive substance — An agent that induces a compulsive craving for its use. Addiction can be physical, psychological, or both.

Algicide — An agent used to control algal blooms.

Anaerobic — A process that does not require oxygen. The production of methane gas is an anaerobic digestion process.

Antagonistic — The process whereby the mixing of two poisons produces a substance less toxic than either substance alone.

Anticholinesterase substance — A substance that destroys the enzyme cholinesterase. Acetylcholine accumulates in the body, causing central nervous system problems.

Antidote — A substance given to neutralize the effects of a poison.

Antimetabolite — A substance that mimics the action of a necessary nutrient in the body. Because the body uses the substance in place of the actual nutrient, growth and bodily processes are inhibited.

Asphyxiant — A substance that crowds out oxygen needed for respiration.

Bacteria — Simple one-celled organisms with naked DNA in the cell. There is protoplasm, but no defined cell nucleus.

Base — A substance that increases the hydroxide concentration of a solution. A proton acceptor.

Blastocyst — The first stage of fetal growth and implantation into the endometrial lining of the uterine wall.

Carcinogen — A substance that induces cancer, which is an uncontrolled growth of cells.

Caustic — A strong base, such as sodium or potassium hydroxide.

Chelation — The use of a substance to combine with, and neutralize the effects of, a metal in the body.

Chelation agent — An antidote used to complex and neutralize the effects of a metal in the body.

Chromatid break — Only one chromatid, or half of the DNA double helix, is broken.

Coagulant — A chemical agent used to precipitate colloidal solids out of raw drinking water or wastewater. Alum is a coagulant.

Coagulant aid — A secondary chemical used to enhance the effect of a primary coagulant. A strengthening agent for floc formation. Polymers are coagulant aids.

Competition — The preferential attraction to a reactant of one chemical over another chemical. Carbon monoxide has a bond attraction 300 times that of oxygen for hemoglobin.

Cross-sensitization — Developing allergies to related substances after exposure to just one substance. An example is that becoming allergic to toluene might predispose a person to react to a first exposure to zylene. Both belong to the family of aromatic compounds.

Cumulative — The combined effects of many exposures to small amounts of a toxin. Exposure to radiation or mercury accumulates damage in the body.

Debride — To remove contamination or irreversibly damaged tissue from a wound in order to enhance healing.

Deliquescent — The tendency of a chemical to absorb moisture from humidity in the air.

Encephalopathy — A disease having to do with the brain. Some protozoa and bacteria will directly attack brain tissue.

Enteric — Having to do with the digestive system. Many waterborne diseases directly upset the stomach and the intestines.

Enzyme inhibitor — A substance that deactivates or destroys an enzyme. Organophosphates destroy cholinesterase.

Epidemiologist — A specialist trained in the tracking of environmentally induced diseases within the community.

Exothermic — Generating heat. Mixing sodium hydroxide with water is exothermic.

Flammable — Capable of catching fire.

Fungi — Multicellular plants incapable of producing their own food. They lack chlorophyll. Fungi may attack living or nonliving matter for the needed nutrients.

Gene — A unit of DNA that controls one hereditary characteristic. This is a codon of three nucleotides.

Granuloma — Tough fibrous tumors and nodules. These can form in the lung after the inhalation of beryllium.

Half-life — The time it takes for one half of an amount of substance to decay, decompose, or be excreted. The biological half-life of mercury is about 70 d. The decay half-life of radium is 1600 years.

Helmiths — Organisms of the tapeworm group.

Hemoglobin — The iron-containing component of blood responsible for the transport of oxygen throughout the body.

Herbicide — An agent used to destroy unwanted vegetation.

Hypersensitivity — An enhanced reaction to exposure to foreign substances. Allergies.

Inhibitor — A substance that delays or stops a chemical or enzymatic reaction.

Irritant — A substance that causes distress upon exposure.

Isochromatid break — An injury to genetic material, involving both parts of the chromosome. Since both parts of the double helix are broken, repair is unlikely.

Meiosis — A cellular process whereby the chromosome number in a nucleus is reduced from 2n to n for the production of gametes.

Microwave radiation — Short radio waves found in radar and certain kinds of digesters. Frequencies between 30 and 300,000 megahertz (MHz).

Mitosis — The fission of a cell into two cells, with replication of the DNA. The chromosome number remains 2n.

Mutagen — A substance causing direct genetic damage. Since cells are damaged at meiosis, the defect is carried to future generations.

Mutagenetic or mutagenic — Causing mutation.

Nematode — Of the roundworm group. Can be free living or parasitic.

Oxidant — An electron acceptor. A substance that increases the oxidation number of another reactant.

Parasite — An organism that lives off another live organism, with benefit only to itself.

Particle radiation — High energy atomic particles capable of interacting with matter.

Pathology — The course of a disease.

Pesticide — An agent used to kill vermin.

Physiology — Having to do with organs and their functions.

Protozoa — The lowest form of animal life having a well-defined cellular nucleus.

Reductant — An electron donor. An agent that reduces the oxidation number of a reactant.

SCBA (self-contained breathing apparatus) — An air supply unit worn in an atmosphere of inadequate oxygen, or noxious gases, for respiratory protection.

SCUBA (self-contained underwater breathing apparatus) — An air supply unit worn for work or sport in an aquatic environment.

Synergistic — The action whereby the combination of two toxins is more toxic that either poison alone.

Teratogen — An agent capable of causing developmental damage without altering the genetic material of the organism. A birth injury agent.

Teratogenetic — Capable of tissue injury without attending DNA damage.

Threshold dose — The dose at which an effect is first perceived.

THM (trihalomethane) — Compounds of halogenated methane. Chloroform is a trihalomethane.

TLV (threshold limit value) — A level of toxin at which effects are first perceived.

Toxicology — A field of study that deals with understanding the action of toxic substances on the body.

TWA (time-weighted average) — The time-weighted average for a normal 8-h workday or 40-h week. Sometimes the literature speaks of a TLV-TWA.

Vaccine — An agent that induces immunity to a disease

Virus — A small obligate parasite of other cells. It has no cellular parts to speak of. It is simply a coat of protein surrounding genetic material of DNA or RNA. It must use another cell for its replication.

Bibliography

1. Anderson, A. 1982. Neurotoxic Follies. *Psychol. Today*. July.
2. American Water Works Association, Inc. 1971. *Water Quality and Treatment: a Handbook of Public Water Supplies*. McGraw-Hill, New York.
3. Babich, H. 1985. Reproductive and carcinogenic health risks to hospital personnel from chemical exposure — a literature review. *J. Environ. Health*. 48(2).
4. Beaconsfield, P., Birdwood, and R. Beaconsfield. 1982. The placenta. *Sci. Am*. 243.
5. Berne, R., and M.N. Levy. 1988. *Physiology*. C.V. Mosby, St. Louis.
6. Cameron, D., and G.A. Hartson. 1988. Aluminum and fluoride in the water supply and their removal for haemodialysis. *Sci. Total Environ*. 76.
7. Carson, J.L., A.M. Collier, Shih-Chin Hu, C.A. Smith, and P. Stewart. 1987. The appearance of compound cilia in the nasal mucosa of normal human subjects following acute, in vivo exposure to sulfur dioxide. *Environ. Res*. 42.
8. Cember, H. 1976. *Introduction to Health Physics*. Pergamon Press, New York.
9. Clark, C. S. 1987. Potential and actual biological related health risks of wastewater industry employment, *J. WPCF*. 59(12).
10. Colwell, R.R., Ed. 1982. *Microbial Hazards of Diving in Polluted Waters*. Maryland Sea Grant Publication No. UM-SG-TS-82-01. University of Maryland, College Park, MD.
11. Stich, H.F., Ed. 1985. *CRC Carcinogens and Mutagens in the Environment: The Workplace*, Vol. 5. CRC Press, Boca Raton, FL.
12. Steere, N.V., Ed. 1971. *CRC Handbook of Laboratory Safety*. 2nd ed. CRC Press, Boca Raton, FL.
13. De Vinck, C. 1985. The power of my powerless brother, *Reader's Dig*. July.

14. Duncan, R.C., and J. Griffith. 1985. Monitoring study of urinary metabolites anad selected symptomatology among Florida citrus workers, *J. Toxicol. Environ. Health*. 16.
15. Fetter, C.W. 1980. *Applied Hydrogeology*. C.E. Merrill Publishing, Columbus, OH.
16. Fradkin, L. et al. 1989. Municipal wastewater sludge: the potential public health impacts of common pathogens. *J. Environ. Health*. 51(3).
17. Gerardi, M.H., A.P. Maczuga, and M.C. Zimmerman. 1988. An operator's guide to wastewater viruses. *Public Works*. April.
18. Gibbons, J.H. 1986. *Reproductive Health Hazards in the Workplace*. 99th Congress. U.S. Office of Technology Assessment, Washington, D.C.
19. Goodmen, E. 1991. Quest for equality in the workplace (editorial). *Akron Beacon Journal*. March 28. Akron, OH.
20. Haley, T.J. 1987. Toluene: a chemical review. *Dangerous Properties of Industrial Materials Report*. September/October.
21. Hammer, M.T. 1975. *Water and Wastewater Technology*, John Wiley & Sons, New York.
22. Hathaway, J.A. 1989. Role of epidemologic studies in evaluating the carcinogenicity of chromium compounds. *Sci. Total Environ*. 86(1-2).
23. Holvey, D.N. and T.J. Talbott, Eds. 1972. *The Merck Manual of Diagnosis and Therapy*, 12th ed, Merck, Rahway, NJ.
24. IOCU. 1986. *The Pesticide Handbook: Profiles for Action*. Organization of Consumer's Union, Penang Malaysia.
25. Jueneman, F.B. 1983. A lead-pipe cinch. *Ind. Res. Dev*. July.
26. Kennedy, W.P. 1967. *Birth Defects Series: Epidemiologic Aspects of the Problem of Congenital Malformations*, Vol. 3. No. 2. National Foundation of March of Dimes, December.
27. Kirschmann, J.D. 1975. *Nutrition Almanac*. McGraw-Hill, New York.
28. Kudo, R.R. 1971. *Protozoology*, 5th ed. Charles C. Thomas, Springfield, IL. January.
29. Krzel, R.B., and C.L. Cetrulo. 1981. The effect of environmental pollutants on human reproduction, including birth defects. *Environ. Sci. Technol*. 15(6), June.
30. Layton, D.W., and R.T. Cederwall. 1986. Assessing and managing the risks of accidental releases of hazardous gas: a case study of natural gas wells contaminated with hydrogen sulfide. *Environ. Int*. 12.
31. Linn, W.S., D.A. Ficher, D.A. Shamoo, C.E. Spier, L.M. Valencia, U.T. Anzar, and J.D. Hackney. 1985. Controlled exposures of volunteers with chronic obstructive pulmonary disease to sulfur dioxide. *Environ. Res*. 37.

32. Lippmann, M. 1990. Lead and human health: background and recent findings. *Environ. Res.* 51.
33. Machenthun, K.M. 1969. *The Practice of Water Pollution Biology.* Federal Water Pollution Control Administration. U.S. Dept. Interior-U.S. Gov. Printing Office, Washington, D.C.
34. New York State Department of Health. 1960. *Manual of Instruction for Sewage Treatment Plant Operators.* Health, Education Service, Albany, NY.
35. Windholz, M., S. Budavari, R.F., Blumetti, and E.S. Otterbein. *Merck Index,* 10th ed. Merck & Co., Inc., Rahway, NJ.
36. Moore, L.L., and E.C. Ogradnik. 1986. Occupational exposure to formaldehyde in mortuaries. *J. Environ. Health.* 49(1).
37. Munley, A.J., R. Railton, W.M. Gray, and K.B. Carter. 1986. Exposure of midwives to nitrous oxide in four hospitals. *Br. Med. J.* 293(25).
38. Nriagu, J.O. 1983. Saturine gout among Roman aristocrates: did lead poisoning contribute to the fall of the empire? *N. Engl. J. Med.* 308(11).
39. Pelczar, M.J., R.D. Reid, and E.C.S. Chan. 1977. *Microbiology.* McGraw-Hill, New York.
40. Hark, E.D., Jr. 1981. Human pulmonary adaptation to ozone. Proceedings of the Research Planning Workshop on Health Effects of Oxidants. EPA-600/9-81-001. January.
41. Proctor, N.H., and J.P. Hughes. 1978. *Chemical Hazards of the Workplace.* J.B. Lippencoot, Philadelphia.
42. Revkin, A. 1983. Paaraquat a potent weed killer is killing people. *Sci. Dig.* June.
43. Savage, E.P. et al. 1988. Chronic neurological sequelae of acute organophosphate pestiscide poisoning. *Arch. Environ. Health.* 43(1).
44. Sawyer, C.N., and P.L. McCarty. 1978. *Chemistry for Environmental Engineering.* McGraw-Hill, New York.
45. Schumann, D. 1990. Nitrous oxide anaesthesia: risks to health personnel. *Int. Nursing Rev.* 37(1).
46. Standeven, A.M., and K.E. Wetterhahn. 1989. Chromium (VI) toxicity: uptake, reduction, and DNA damage. *J. Am. Coll. Toxicol.* 8(7).
47. Tilton, B.E. 1989. Health effects of tropospheric ozone, *Environ. Sci. Technol.* 23(3).
48. University of South Florida. 1984. *Pesticides in Ground Water Symposium.* Tampa, FL. May 10.
49. Ups and Downs of Discrimination, *U.S. News,* April 1, 1991.
50. Vaughn, T.M., E.F. Landry, and M.Z. Thomas. 1983. Entrainment of viruses from septic tank leach fields through a shallow, sandy soil aquifer. *Appl. Environ. Microbiol.* 45(5).

51. Waldbott, G.L. 1978. *Health Effects of Environmental Pollutants*. C.V. Mosby, St. Louis.
52. Joint Committee of the Water Pollution Control Federation and American Society of Civil Engineers. *Wastewater Treatment Plant Design*. 1982. Lancaster Press, Lancaster, PA.
53. Weber, W., Jr. 1972. *Physiochemical processing for watar quality control*. John Wiley & Sons, New York.
54. Weiner, M.A. 1987. *Reducing the Risk of Alzheimer's*. Stein and Day, Briarcliff Manor, NY.
55. Willard, H.H., L.L. Merritt, and J.A. Dean. 1974. *Instrumental Methods of Analysis*. D. Van Nostrand, New York.
56. Wilsone, J.S. 1981. The Knowledge and Behavior of Pregnant Women in Relation to Teratogenic Agents and Teratogenic Behaviors. Master's Thesis. Kent State University, Kent, OH.
57. Zayed, J. et al. 1990. Environmental contamination by metals and Parkinson's disease. *Water Air and Soil Pollut.* 49.

Index

A

AA, see Atomic absorption
Acetylcholinesterase, 42
Acids, see also specific types
  as pulmonary irritants, 61
  first aid for injury, 61
  self-protection, 62, 63
  spill containment, 62, 63
  storage of, 62
  ventilation practices, 62
Activated carbon, 74
Addictive agents
  acute poisoning symptoms, 49
  chemical composition, 49
  defined, 48
  effects on CNS, 48, 49
  first aid, 49, 51
  rehabilitation, 51
  self-protection, 49
  storage, 49
  toluene, see Toluene
  turpentine, see Turpentine
  ventilation practices, 49
  xylene, see Xylene
Adenoviruses, 87
Aeromonas, 91
Agent Orange, 101
AIDS, 86, 89
$Al_2(SO_4)_3 \cdot 14H_2O$, see Alum
Alcohols, see specific types
Alkali earth bases, see specific types
Allergic reactions, 49
Alum ($Al_2(SO_4)_3 \cdot 14H_2O$)
  acute poisoning symptoms, 35
  defined, 35
  disease occurrence, 36
  first aid, 35, 37
  human daily intake, 36
  impact on diet, 36
  long-term effects, 35
  jar testing, 35
  pathology, 37
  precautions, 37
  self-protection, 36
  source, 37
  storage, 35
  TLVs, 35
  ventilation practices, 36
Ammonia (NH_3)
  acute poisoning symptoms, 18
  first aid, 18
  properties/recommendations, 21
  skin activity, 18
  TLVs, 17
  ventilation practices, 18
Amyotrophic lateral sclerosis, 36
Anaerobic processes, 9, 10
Anemia, 29, 36, 49
Antagonistic effects, of chemical mixtures, 118
Aromatics, 49
Artificial respiration, 8, 15
Ascaris lumboricordes, 89
Aspergillus
  flavus, 88
  fumigatus, 88

niger, 88
Asphixiants, see specific topics
Asthma, 15–17, 58
Astroviruses, 86
Atomic absorption (AA), 11, 12, 23, 51
Atropine sulfate, 44

B

Bacteria, see also specific types
　immunization against, 87–88
　sewage as growth medium for, 88
BAL, 25
Balantidium coli, 88
Bases
　as protein dissolvers, 67
　first aid, 67
　self-protection, 67
　spill containment, 67
　strong
　　KOH, see Potassium hydroxide
　　NaOH, see Sodium hydroxide
　type of burns received from, 67
　ventilation practices, 67
　weak
　　CaO, see Calcium oxide
　　Ca(OH)$_2$, see Calcium hydroxide
　　Na$_2$CO$_3$, see Sodium carbonate
　　NaHCO$_3$, see Sodium bicarbonate
Be, see Beryllium
Benzene, 58
　defined, 57
　first aid, 57
　laboratory safety procedures, 58
　organ damage, 57
　OSHA safety standards, 57
　TLVs, 57
　toxicity levels, 57
　ventilation practices, 58
Beryllium (Be)
　acute poisoning, 31
　defined, 31
　first aid, 31, 37
　pathology, 37
　precautions, 37
　source, 37
　self-protection, 31
　skin activity, 31
　TLVs, 31
　ventilation practices, 31
Bipyridyl compounds, 47
　acute poisoning symptoms, 48
　defined, 47
　first aid, 48
　laboratory safety procedures, 48
　self-protection, 48
　skin activity, 48
　TLVs, 48
　toxicity levels, 48
Birth defects, 42; see also Human reproduction
　laboratory safety standards, 104–105
Blue vitriol, 34
Bronchitis, 17, 64
Burns
　categories of, 76–77
　death
　　from infection, 76
　　from shock, 76
　debriding, 76
　treatment of, 76–77

C

Cadmium (Cd)
　first aid, 30, 38
　long-term effects, 30
　defined, 30
　"itai-itai" bone disease, 30

INDEX

pathology, 38
precautions, 38
source, 38
TLVs, 30
ventilation practices, 30
Calcium hydroxide, 67
  application, 70
  first aid, 70
  precautions, 70
  reactivity, 70
Calcium oxide (CaO), see also Quicklime
  application, 70
  cooling provisions, 68
  first aid, 70
  precautions, 70
  reactivity, 70
  self-protection, 68
  storage, 68
  use of, 68
  violent reactions with, 68
CaO, see Calcium oxide
Ca(OH)$_2$, see Calcium hydroxide
Carbon dioxide (CO$_2$)
  acute poisoning symptoms, 6
  defined, 6
  first aid, 6
  instruments for detection
    gas-ratio analyzer, 6
  properties/recommendations, 20
  ventilation practices, 6, 8
Carbon monoxide (CO)
  acute poisoning symptoms, 8; see also specific types
  defined, 8
  effects on pregnancy, 9, 100
  first aid, 8
  placental respiratory insufficiency, cause of, 94
  properties/recommendations, 20
  seasonal precautions, 9
  TLVs, 8
  ventilation practices, 9
Carboxyhemoglobin, effects on exposed persons, 8
Cardiopulmonary respiration (CPR), 1, 8, 15, 49
Cartridge respirators, 6, 14
Caustics, see specific types
Cd, see Cadmium
Central nervous system (CNS), effects on
  by CO, 8
  by fluoride, 66
  by H$_2$S, 11
  by Hg, 24
  by Mn, 30
  by N$_2$O, 12
  by organics, 48
  by Pb, 28
  by phenol, 56
  by systemic poisons, 55–59
  general asphixiants effects, 5
Cestodes, 89
Chelating agents, 25, 28, 36
Chlorine (Cl$_2$)
  acute poisoning symptoms, 15
  properties/recommendations, 20
  TLVs, 13
  ventilation practices, 15
Chlorine dioxide (ClO$_2$)
  acute poisoning symptoms, 15
  manufacture of, 13
  premature aging, 15
  properties/recommendations, 20
  TLVs, 15
  tooth erosion, 15
  ventilation practices, 15
Chlorine gas, 14, 16
Chlorites, 74
Chloroform, see Halogenated hydrocarbons
Cholera, 87
Cholinesterase enzyme, 44
Chromate salts, 75, 80–81

Chromerge®, 32
Chromium (Cr)
  acute poisoning symptoms, 33
  defined, 31
  effect on human reproduction, 98
  first aid, 33, 38
  human daily intake, 31
  laboratory safety procedures, 32
  OSHA safety standard, 31
  oxidation states, 31–32
  pathology, 38
  precautions, 38
  self-protection, 32
  source, 38
  storage, 32–33
Cigarette smokers, susceptibility to toxic materials, 8, 16, 95
Cl_2, see Chlorine
ClO_2, see Chlorine dioxide
Clostridium teteni, 87–88
CNS, see Central nervous system
CO, see Carbon monoxide
CO_2, see Carbon dioxide
Coagulants, see specific types
Cobalt 60, 101
Cold-vapor analysis, 99
Compulsive obsessive behavior, 30
Contact dermatitis, 31, 33
Contact lenses, dangers of wearing when working with chemicals, 18, 61
Copper sulfate ($CuSO_4$)
  acute poisoning symptoms, 34
  defined, 33–34
  first aid, 34, 38
  human daily intake, 34
  long-term effects, 34
  pathology, 38
  precautions, 38
  self-protection, 34
  skin active, 34
  source, 38
  storage, 34
  TLVs, 34
Corneal damage, 11, 18, 59, 67
Corrosives, see specific types
CPR, see Cardiopulmonary respiration
Cr, see Chromium
CRC Handbook of Laboratory Safety, 80, 82, 107
Cross-sensitization, 49, 118
Crown corrosion damage, 10, 11
Cryptosporidium, 88
$CuSO_4$, see Copper sulfate

D

Depression, 44, 52
Dermatitis, 47
Dialysis encephelopathy, 36
Dichloramines, 17
Dichromate salts, 75, 80–81
Dimercoprol, 25
Disease agents, 91–92
Diving, precautions, 90–91
Dizziness, 8, 44, 49, 54

E

E. coli, 87
Edema, pulmonary, 115
  from Cl_2, 13
  from ClO_2, 13
  from H_2S, 10, 11
  from NH_3, 18
  from NO, 62
  from NO_2, 62
  from O_3, 19
  from SO_2, 17
  from sodium chlorite byproducts, 74
Electron capture detectors, 103
Emphysema, 15, 17, 19

Encephalitis, primary amoebic, 91
Entamoeba histolytica, 88
Environmental contamination, 3
Enzymes, 42
Ether
 fuel-to-oxidant ratio, 81
 low flash points, 81
 peroxides, 75, 81
  testing for presence of, 81
 shelf life of, 81
 storage of, 81
 ventilation practices, 81
Ethyl alcohol, 53–54
Ethyl ether, 51–53, 81
 acute poisoning symptoms, 52
 defined, 52
 first aid, 52
 laboratory dangers of, 81
 peroxides, formation of, 52
 skin activity, 52
 storage, 53
 TLVs, 52
 use in AA, GC, 52
 ventilation practices, 53

F

$FeCl_2$, see Ferric chloride
Ferric chloride ($FeCl_2$)
 acute poisoning symptoms, 34
 defined, 34
 first aid, 35, 36, 38
 human daily intake, 34
 pathology, 38
 precautions, 38
 skin active, 35
 source, 38
 storage, 34–35
 TLVs, 35
Fetus, damage to, see Human reproduction
Fibrosis, 48
Field chemical hazards, 33–38; see specific topics

Fire blankets, 76, 77
Fire
 classes of, 79
 containment of, 77, 80
 extinguishers, types of, 77–79
First aid, 1, 2
Fluoride
 chizzola maculae, 66
 delayed onset of acute poisoning symptoms, 66
 effects on CNS, 66
 fluorosis, 66
 human daily intake, 66
 laboratory safety procedures, 66
 long-term effects, 65
 mimicking symptoms, 66
 storage, 65
Fluospar, as water fluoridation substitute, 65
Formaldehyde gas, see Formalin
Formalin, 58
 acute poisoning symptoms, 59
 carcinogenic agent, 58
 cross-sensitization, 58
 defined, 57
 first aid, 59
 self-protection, 59
 TLVs, 58
 ventilation practices, 59
Freon, see Halogenated hydrocarbons
Fuel-to-oxidant ratios, 12, 24
Fungi, see specific types
Fungal infections, susceptibility to in AIDS, 89

G

Gas chromatography, 41, 55, 101, 103
Gasoline, 48, 49
Genetic material, damage of, 95–96
Giardida lamblia, 88

"Glue heads", 51
Gout, 29
Gram-negative infections, 87, 91
Gram-positive infections, 87
Granulomatous inflammation, 31

H

H_3PO_4
  applications, 69
  first aid, 69
  precautions, 69
  reactivity, 69
H_2SO_4, see Sulfuric acid
Halogenated hydrocarbons; see also specific types
  acute poisoning symptoms, 55
  carcinogenic effects, 55
  first aid, 55
  organ damage, 55
  self-protection, 55
  TLVs, 55
  ventilation practices, 55
Hay fever, 47
HCl, see Hydrochloric acid
$HClO_4$, see Perchloric acid
Headache, 8, 54
Heavy metals, 23–38
  algicides
    copper sulfate, see Copper sulfate
  coagulants
    alum, see Alum
    ferric chloride, see Ferric chloride
  laboratory/plant sources of contact
    beryllium, see Beryllium
    cadmium, see Cadmium
    chromium, see Chromium
    lead, see Lead
    manganese, see Manganese
    mercury, see Mercury
  skin active, 33

Hemoglobin, 8, 9
Hepatitis, 86
Hexane, 53
HF, see Hydrofluoric acid
Hg, see Mercury
HNO_3, see Nitric acid
Human reproduction, 93–105
  birth defects, 93, 97
  fetal damage, 94, 95, 98–99, 102
  laboratory hazards, see specific topics
  men, 93
    CNS damage, 98
    sperm damage, 98, 100, 102
    mental retardation, 94, 103
    mutagenic agents, 94
    teratogenic agents, 94
  women
    CNS damage, 98
    discrimination against, 93, 94
    menstrual cycle disruption, 98
    miscarriages, 98, 100, 103
    ovary damage, 98
    pregnancy, 94, 98
Hydrochloric acid (HCl)
  applications, 69
  first aid, 62, 69
  fuming problems with, 62
  delayed symptoms of lung damage with use, 62
  precautions, 69
  reactivity, 69
  TLVs, 62
  ventilation practices, 62
Hydrofluoric acid (HF)
  applications, 69
  corrosive properties of, 65
  defined, 64
  first aid, 65, 69
  nerve injury, 65
  precautions, 69
  reactivity, 69
  skin activity, 65

INDEX   135

water fluoridation substitutes, 65
Hydrofluosilicic acid, as water fluoridation substitute, 65
Hydrogen sulfide (H_2S)
  acute poisoning symptoms, 10
  effects on human reproduction, 100
  first aid, 11
  instruments for detection, 10
  long-term effects, 11
  properties/recommendations, 20
  skin activity, 10, 11
  TLVs, 10
  ventilation practices, 11
Hygiene, in workplace, 90
Hypochlorites, 74

I

Immunization, 89–90, 104
Impaired judgment, 8
Infrared detection, 41
Irritability, 8, 44
Isopropyl alcohol, 53–54

J

Jar testing, 23, 35

K

Kjeldahl nitrogen, 23
Klebsiella, 91
KMnO, see Potassium permanganate
KOH, see Potassium hydroxide

L

Laboratory chemical hazards, 23–33; see specific topics
Laboratory ventilation practices, 23–24
Lead (Pb)
  acute poisoning symptoms, 27, 29
  defined, 27
  effect on human reproduction, 98
  first aid, 29, 38
  human daily intake, 29
  long-term effects, 27–29
  in paint, 97
  pathology, 38
  poisoning, see Plumbism
  precautions, 38
  source, 38
  TLVs, 97
  ventilation practices, 29
Legionella pneumophila, 87
Leptospirosis icterohaemaorrhagiae, 87
Limes, see specific types
Liquid chromatography, 41
"Lockjaw", see *Clostridium teteni*

M

"Mad Hatter" syndrome, 24
Manganese (Mn)
  acute poisoning symptoms, 30; see also specific types
  defined, 29
  first aid, 29, 38
  human daily intake, 29
  pathology, 38
  precautions, 38
  self-protection, 29
  skin active, 29
  source, 38
  storage, 30
  TLVs, 29
  ventilation practices, 30
Marigold flowers, 47
Material Safety Data Sheet (MSDS), 2, 75, 96, 116, 118
Medulla, effect of toxicity on, 10
Mental confusion, 8

Mental retardation, 94
Mercury (Hg)
  acute poisoning symptoms, 24
  carbon trap, , 25, 26
  cold-vapor analysis, fumes from, 25
  damage to genetic material from, 98
  defined, 24
  effects on pregnancy, 24
  first aid, 25, 38
  fume control, 26, 28
  long-term effects, 24
  "Mad Hatter" syndrome, 24
  paint preservative, 98
  pathology, 38
  precautions, 38
  self-protection, 25, 99
  sickness, 25
  source, 38
  TKN analysis, fumes from, 25
  TLVs, 24, 98
  ventilation practices, 25, 99
Metabolic acidosis, 51
Metabolism disrupters, 47–48
Metal corrosion, sources of, 10, 11, 35
Metal-fume fever ("Welder's fever"), 23, 27
Methane (CH_4)
  defined, 9
  first aid, 10
  instruments for detection
    gas ratio analyzer, 6
  properties/recommendations, 20
  ventilation practices, 9, 10
Methyl alcohol, 53–54
Methyl carbamates, 45–46
Methyl chloride, see Halogenated hydrocarbons
Methyl isobutyl ketone
  acute poisoning symptoms, 52
  defined, 51
  first aid, 52
  TLVs, 51
  toxicity levels, 51
  use in AA, GC, 51
  ventilation practices, 51
Microbial hazards, in water, 85–92
  bacteria, 87–88
  diseases from, 85; see also specific diseases
  effects on human reproduction, 103–104
  fungi, 88–89
  helminths, 89
  protozoa, 88
  routes of transmission, 85
  viruses, 86–87
Microwave radiation
  effects on pregnancy, 102
  from ovens, 102
  leakage, 102
  sperm damage, 102
Mn, see Manganese
Monochloramines, 17
"Montezuma's revenge", see *Entamoeba histolytica*
Moodiness, 29
MSDS, see Material Safety Data Sheet
Mutagens, 94; see also specific types

N

Na_2CO_3, see Sodium carbonate
$Na_2Cr_2O_7$, see Sodium dichromate
Naegleria fowleri, 91
NaF, see Sodium fluoride
$NaHCO_3$, see Sodium bicarbonate
NaOH, see Sodium hydroxide
Na_2SiF_6, see Sodium silicofluoride
Nematodes, 89
Nerve agents
  methyl carbamates, see Methyl carbamates

organophosphates, see Organophosphates
pyrethrum-based insecticides, see Pyrethrums
Neutralization, 2, 59
NH_3, see Ammonia
Nickel
　acute poisoning symptoms, 97
　as laboratory hazard, 97
　defined, 97
　self-protection, 97
　TLVs, 97
　ventilation practices, 97
Nitric acid (HNO_3), 75
　applications, 69
　dangerous byproducts of, 62
　first aid, 64, 69
　precautions, 69
　properties of, 62, 64, 80
　reactivity, 69
　self-protection, 62
　spill containment, 62, 63, 80
　storage of, 62, 64
　TLVs, 62
　ventilation practices, 64
　violent reactivity, 62
Nitric oxide (NO), 62
Nitrogen dioxide, 62
Nitrous oxide (N_2O)
　acute poisoning, 12–13
　birth defects, 99
　byproducts, danger of, 13
　CNS, effects on, 12
　first aid, 13
　fuel-to-oxidant ratios, 12
　importance of good nutrition, 100
　NIOSH safety levels, 99
　properties/recommendations, 20
　reversible sperm damage, 99
　role in miscarriages, 99
　TLVs, 12, 13
　tooth erosion, 13
　vitamin B_{12}, destruction of by, 95, 100

NO, see Nitric oxide
NO_2, see Nitrogen dioxide
Norwalk virus, 86

O

Opportunistic feeders, 91
Optic nerve damage, 53
Organics, health effects of, see also specific topics
　acid-reactive, 60
　addictive agents, 42
　carcinogens, 42
　effects on CNS, 59
　oxidant-reactive, 60
　reproductive effects, 42
　skin active, 59
　threat to CNS, 42
Organophosphates, 43
　acute poisoning symptoms, 44
　chemical composition, 43–44
　defined, 43
　effects on CNS, 41
　first aid, 44
　history of, 43
　long-term effects, 44
　malathion, 43
　oxone group, 44
　parathion, 43
　self-protection, 45
　storage, 45
　TLVs, 41
　usage, 45
　ventilation practices, 45
Oxidants, see also specific chemicals, compounds
　dangers in laboratories, 75–83
　　burns, see Burns
　　fire, 77–78
　　fire blankets, 76
　　first aid kits, 76
　　safety program, 78–80
　　storage of chemicals, 75
　dangers in plants, 73–75
　defined, 73
　reactivity, 73

Oxidation-reduction practices, 12
Oxygen, 6
 instruments for detection
  gas ratio analyzer, 6
Ozonation, 19
Ozone (O_3)
 acute poisoning symptoms, 19
 effects on human reproduction, 100
 generator, 19
 premature aging, 19
 properties/recommendations, 21
 TLVs, 100
 ventilation practices, 19

P

Parkinson's disease, 30, 35
Particulate respirators, 6
Pb, see Lead
Perchloric acid ($HClO_4$), 75
 defined, 81
 properties of, 82
 spill containment, 82
 storage of, 82
 use of wash-down hood, 82
Pesticides
 effects on human reproduction, 100–101
 mutagenic/teratogenic effects of, 100
Phenol, 56
 acute poisoning symptoms, 55–56
 chemical composition, 55
 defined, 55
 effect on CNS, 56
 first aid, 57
 laboratory safety procedures, 56–57
 organ damage, 56
 ventilation practices, 57
Placental respiratory insufficiency, 94

Plumbism, 27, 98
Polio, 85, 86
Potassium hydroxide (KOH), 67
 application, 70
 first aid, 70
 precautions, 70
 reactivity, 70
Potassium permanganate ($KMnO_4$), 29, 75, 80–81
 reactivity with activated carbon, 74
 spill containment of, 73–74
 storage of, 73
 use of, 73
Premature aging, 15, 19
Priority pollutants, 48, 101
Protozoa, see specific types
Pseudomonas, 91
Pulmonary agents, 5–21
 asphyxiants, 5–11
 CH_4, 9–10
 CO, 8–9
 CO_2, 5–8
 H_2S, 10–11
 respiratory irritants, oxidants, 11–21
 Cl_2, 11, 13, 15–16
 N_2O, 11–13
 NH_3, 11, 17–18
 O_3, 18–19
 SO_2, 11, 16–17
Pulmonary tumors, 19, 31, 48
Pyrethrums, 46–47
 allergies to, 47

Q

Quicklime, 68

R

Radiation
 ionizing
  effects on human reproduction, 102
  neutron sources, 102

INDEX 139

tolerance levels, 102–103
longer-wave
 high-power electrical fields, 101
 microwave, see Microwave radiation
RBC, see Rotating Biological Contactor
Red blood cell cholinesterase, 45
Reducants, see also specific chemicals, compounds
 dangers in laboratories, 75–83; see Oxidants
 dangers in plants, 73–75
 defined, 73
 reactivity, 73
Removal of victim, 2, 8, 11, 52
Reproductive effects, 42
Resuscitation of victim, 2, 8, 11, 52
Rotating Biological Contactor (RBC), source of CO_2, 6

S

Safety program, 107–113
 emergency preparedness, 108–109
 first aid stations, 109–111
 library, 107
 plan, 112–113
 professional assistance, 109–110, 112
 training, 108
 ventilation practices, 109
Salmonella typhi, 87
SCBA, see Self-contained breathing apparatus
Self protection, 2
Self-contained breathing apparatus (SCBA), 6, 7, 15, 16, 19, 80
Sevin® dust, 45
Shortwave-frequency radiation, see Microwave radiation

Sickle-cell anemia, 9
Skin active chemicals, 115; see also specific chemicals
Skin active gases, 10, 18, 115
SO_2, see Sulfur dioxide
Sodium bicarbonate ($NaHCO_3$), 67
 application, 70
 first aid, 70
 precautions, 70
 reactivity, 70
Sodium carbonate (Na_2CO_3), 67
 application, 70
 first aid, 70
 precautions, 70
 reactivity, 70
Sodium chlorite, 13, 74
Sodium chromate salts, 74
Sodium dichromate ($Na_2Cr_2O_7$), 74
Sodium fluoride (NaF)
 applications, 70
 as water fluoridation susbstitute, 65
 first aid, 70
 precautions, 70
 reactivity, 70
Sodium hydroxide (NaOH)
 application, 70
 first aid, 70
 precautions, 70
 reactivity, 70
 self-protection, 68
 storage of, 68
 use of, 68
 ventilation practices, 68
Sodium silico-fluoride (Na_2SiF_6)
 applications, 69, 70
 as water fluoridation substitute, 65
 first aid, 69, 70
 precautions, 69, 70
 reactivity, 69, 70
Soldering, 23, 27
Solvents, exposure to, 48
Sulfur dioxide (SO_2)

acute poisoning symptoms, 17
allergy to, 17
dechlorination, 16
first aid, 17
properties/recommendations, 20
TLVs, 16
ventilation practices, 16
Sulfuric acid (H_2SO_4), 62
applications, 69
first aid, 69
precautions, 69
reactivity, 69
Swimmer's ear, see *Aspergillus*
Synergistic effects, of chemical mixtures, 118
Systemic poisons, see also Benzene; Formalin; Phenol
acute poisoning symptoms, 55
defined, 55
effect on CNS, 55
first aid, 55
self-protection, 55
ventilation practices, 55

T

Teratogens, 94; see also specific types
Tetanus, 85, 87–88
Threshold Limit Values (TLVs), 117; see also under specific chemicals
TKN, see Total Kjeldahl nitrogen analysis
TMHs, see Trihalomethanes
Toluene, 50
 acute poisoning symptoms, 49
 addiction to, 51
 defined, 49
 first aid, 49, 51
 metabolic acidosis, 51
 rehabilitation, 51
 self-protection, 49
 TLVs, 49
 toxicity levels, 49
 ventilation practices, 49
Tooth erosion, 13, 15, 64
Total Kjeldahl nitrogen analysis (TKN), 23, 99
Trichloramines, 17
Trihalomethanes (TMHs), 19
Tuberculosis, 15
Turpentine, 50
 addiction, 51
 allergic reactions to, 49
 effects on CNS, 49
 first aid, 49, 51
 kidney damage from, 49
 rehabilitation, 51
 self-protection, 49
 TLVs, 49
 ventilation practices, 49
Typhoid, 85, 87, 90

V

Vibrio cholerae, 87
Viruses, see specific types
Vitamin B_{12}, 12, 95, 100
Vitamin C, 29

W

Water reclamation, 1
Weakness, 8, 11, 30, 57
"Welder's fever", see Metal-fume fever
Wound flushing, 2

X

Xylene, 49